MATH MADE A BIT EASIER

MATH MADE A BIT EASIER

Basic Math Explained in Plain English

LARRY ZAFRAN

Self-published by author via CreateSpace
Available for purchase exclusively on Amazon.com

MATH MADE A BIT EASIER:
Basic Math Explained in Plain English

Self published by author via CreateSpace
Available for purchase exclusively on Amazon.com

Book design by Larry Zafran

Printed in the United States of America
First Edition printing November 2009

ISBN-10: 1-4495-6510-7
ISBN-13: 978-1-44-956510-7

Please visit the companion website below for additional information, to ask questions about the material, to leave feedback, or to contact the author for any purpose.

www.MathWithLarry.com

CONTENTS

CHAPTER ZERO 7
Introduction

CHAPTER ONE 29
Is Math Hard, and If So, Why?

CHAPTER TWO 51
The Foundation of Math:
Basic Skills in Arithmetic

CHAPTER THREE 77
Basic Math Topics and Operations

CHAPTER FOUR 103
Working with Negative Numbers

CHAPTER FIVE 123
Basic Operations with Fractions (+, –, ×, ÷)

CHAPTER SIX 133
More About Fractions

CHAPTER SEVEN 159
Other Topics in Fractions

CHAPTER EIGHT 175
The Metric System, Unit Conversion,
Proportions, Rates, Ratios, Scale

CHAPTER NINE 195
Working with Decimals

CHAPTER NINE AND FIVE-TENTHS 213
More Topics in Decimals

CHAPTER TEN 229
Working with Percents

CHAPTER ELEVEN 249
Basic Probability and Statistics

CHAPTER TWELVE 263
How to Study and Learn Math,
and Improve Scores on Exams

About the Author & Companion Website 279

CHAPTER ZERO

INTRODUCTION

A DIFFERENT PERSPECTIVE ON MATH EDUCATION

This book was written based on my experience as an independent math tutor, and as a former school teacher of math who changed careers from an unrelated industry. With my intent to independently self-publish the book, my writing was not bound by the requirements or guidelines of any major publisher who would likely be uncomfortable with its non-mainstream, unique style.

Because of my independence, I am in a position in which I risk nothing by telling the unpleasant truth about why students struggle with math, as well as what can be done to improve the situation. This is not a generic textbook. Every word is based on the patterns that I've personally, repeatedly observed among many students of all ages.

THE PURPOSE OF THIS BOOK

In my independent tutoring work, I continually observe that almost all of the difficulty that students have with math follows a distinct pattern. They all have very similar complaints, and they all get stuck in almost the exact same places. It is too consistent to be a coincidence.

With that in mind, this book serves several purposes. The first is to explain very clearly why students struggle with math. This book was written from a standpoint that acknowledges how hard math is for most students, but also maintains that there is a reason for why that is the case, and that there is a means of ending the struggle.

This book serves as a carefully planned roadmap for self-studying and learning math. Topics that require extensive review and redundancy receive it, and those that do not are only covered in passing. The connection between topics and their dependence upon one another are very clearly highlighted. Nothing is presented in a vacuum.

This is contrary to most commercial textbooks, test-prep books, and review books which typically devote the same amount of space to every topic. There is no distinction between topics that most students find easy and ones that most students find difficult, nor is there a distinction between trivial topics and very significant ones that need to be repeatedly explained in many different ways, even at the risk of redundancy.

While many math textbooks include embedded textboxes of "tips," I have yet to see a tip which reads, "If you do not fully understand and master this essential concept, you will be absolutely sunk when you study later math." This book makes such statements where applicable and appropriate. They are not intended as threats, but indeed guarantees based on experience and observation.

Finally, this book presents some important advice and suggestions for how to study and learn math, and how to improve scores on exams. They are based on my personal experiences as a student, as well as what has proven successful for my tutoring clients.

Most students never learn how to study effectively, and never learn how to evaluate their own readiness for exams. This book attempts to change that by presenting some ideas that aren't commonly discussed elsewhere.

HOW THIS BOOK IS ORGANIZED

Chapter One of this book acknowledges the fact that most students find math to be extremely hard, and attempts to explain why that is the case, and what to do about it. Chapters Two through Eleven cover subject material starting with concepts in basic arithmetic and operations which are the foundation of all math studies.

It goes on to cover the basics of negative numbers, fractions, decimals, percents, probability, statistics, the Metric system, and measurement.

Chapter Twelve offers suggestions for how to study and learn math, and how to improve scores on exams. The book concludes with information about the companion website which provides additional help and information, as well as the means of asking me questions by e-mail.

GRADE LEVEL OF THE MATERIAL COVERED

This point of concern is very much representative of why students struggle so much with math. Math is not about having a student in a particular grade do math which has been designated as being "on level" for that grade. Math must be learned step-by-step starting from the very beginning, progressing at whatever pace is necessary.

When a student has trouble with an algebra topic, for example, the response is usually to seek out algebra tutoring or an algebra review book. That is almost always the wrong approach because in virtually all cases, it is not algebra which is the source of the problem. The problem is usually that the student never fully mastered the basic math topics upon which the algebra is based.

Stated in math terms, if Johnny can't compute 3 – 5, he can't be expected to simplify $3x^2y^3 - 5x^2y^3$. The problem is rarely that Johnny "gets confused with the *x*," as many believe. Understanding and accepting this is the key to ending the struggle with math.

This book simply starts at "the beginning." There is nothing "babyish" about the material. Until all the topics in the book are fully mastered, there is truly no point in progressing to any later material. It will just be a matter of feeling angry, frustrated, confused, and helpless.

If you want to get better at math, stop thinking about grade levels, and stop thinking about tomorrow's test that you may feel totally unprepared for. Just think about learning math. Admittedly it is hard to think in those terms when you are facing a particular academic deadline, but math will only start to get easier once you do.

ASSUMED MATH KNOWLEDGE

This book assumes that the reader has at least a second grade math level, namely some very basic experience with the way numbers and math work in a general sense. It is rare that this is the problem for students. For most students, the problem begins in the third or fourth grade when the math starts to get somewhat abstract, and where some independent thinking and problem-solving skills become required. This is when students begin falling victim to a education system which "pushes them through" to higher level math without their ever fully grasping the earlier material. Gaps begin to develop, and those gaps quickly widen and get deeper over the years.

This book also assumes that the reader is capable of reading on at least a middle school grade level, and that s/he has the discipline and sophistication to study the material on his/her own. However, I answer students' questions for free via my website, so in a sense you're not as alone as you might think you are.

A reading deficiency can play a huge role in a student's struggle with math. If a student has difficulty reading the words that comprise a word problem, or the words in a problem's instructions, that issue must be fully addressed before the student declares that "math is hard."

Note that this book assumes that the reader has at least seen the material that is being presented, even if it was a long time ago and/or was never fully understood. It is extremely hard to learn totally brand new topics from a book, especially one like this which attempts to include a large amount of a material in relatively little space.

THE OVERALL TONE OF THE BOOK

The overall tone of this book is serious, but with a conversational tone. Students of all ages including adults struggle tremendously with math, and it doesn't help matters to present the material in the form of cutesy cartoons, or humorous, contrived examples and stories which barely relate to the material. Certainly it serves no purpose to refer to the reader as a "dummy," "idiot," or anything similar, even if in a tongue-in-cheek manner.

My goal for this book is that while you read it, you'll feel as though I'm your personal math tutor sitting beside you and answering your questions "in plain English," as promised by the book's subtitle. In writing the book, I drew upon my teaching and tutoring experience to address the questions that most students usually have at given points in the material.

The book does not insult the reader by implying that math is easy, fun, or exciting, nor does it promise to make it any of those things. It also doesn't promise that the reader will learn math "fast," let alone in a specific number of days or weeks. Its goal is just to make math "a bit easier," as indicated in the title. Students who hate math and struggle with it may continue to do so after reading this book, but hopefully to a lesser extent.

This book doesn't claim that a student's struggle with math is his/her fault, but it does emphasize the obvious fact that all the effort to get better at math ultimately must come from the student. It acknowledges the fact that for most students, the study of math is simply a means to an end, and that much of the material has little to no direct relevance in everyday life. It also acknowledges that we live in a world of calculators and computers, and that most exams are designed with this in mind. Some basic arithmetic is covered for its valuable instructional benefit of reinforcing important concepts, but not to unrealistically imply that we live in a world in which cashiers total grocery bills by hand.

INTENTIONAL REDUNDANCY IN THE BOOK

This book is intentionally redundant to the extent that it is required to drive home certain very key points. It is far better to present such concepts repeatedly and in different ways than it is to just mention them in passing.

This is by far one of the biggest problems with traditional textbooks and review books. No author or publishing company wants to risk being accused of "going on and on" about the same thing. This book takes that risk. There is no such thing as over-learning very significant topics. Conversely, there is no point in spending a great deal of time on topics that most students either find very straightforward, or do not serve as important prerequisites (required prior knowledge) for later material.

WHY THIS BOOK DOESN'T EMPHASIZE WORD PROBLEMS

Word problems are intentionally quite sparse in this book. There are several reasons for this. One reason is that is it extremely difficult to teach students how to solve word problems even in a classroom or private tutoring setting, let alone from a book. Most students struggle with word problems primarily because they are not skilled and careful readers. That is something which simply takes practice and hard work over an extended period of time, and cannot be taught from a book.

The only way to get better at word problems is to study all of the underlying math concepts. There are an infinite number of possible word problems that can be posed, and it is often the case that changing one single word results in an entirely new problem. You will have the greatest chance for success if you fully understand all of the math concepts that are you going to be tested on, and if you can discipline yourself to read word problems carefully and accurately.

WHY THIS BOOK DOESN'T EMPHASIZE CHARTS, TABLES, AND GRAPHS

Many commercial test prep books are filled with charts, tables, and graphs. To some extent this is because many students consider themselves to be visual learners, and are more likely to purchase such a book.

The reality, though, is that most students do not have much difficulty with interpreting charts, tables, and graphs. Such problems typically involve nothing more than using common sense. As with word problems, it is difficult to actually teach that skill, let alone by way of a book. Also, it is impossible to prepare for every possible chart, graph, or table that you may encounter.

The best you can do is just make sure that you understand all of the underlying concepts. For the most part, the difficulty that students have with visual problems are usually based on careless errors or lack of careful observation. Again, this is something that cannot be addressed by way of a book.

THE BOOK'S POSITION ON CALCULATOR USE

This book takes a realistic and modern position on calculator use. Calculators are permitted on most standardized exams, and more and more schools and teachers are permitting their use in the classroom. The trend in math education is for students to learn how to think creatively and become proficient at problem solving, as opposed to doing arithmetic by hand.

With that in mind, this book only discusses certain basic arithmetic tasks for the purposes of reviewing important concepts. If you are going to be tested on performing arithmetic operations by hand, and you don't feel confident in your ability to do such, you'll need to find a source to help you. It is very difficult to explain such procedures in a book, so it is suggested that you search for some free video clips on the Internet, or seek out a local tutor. As with anything else, make sure that you get extensive practice, and more importantly make sure that you understand why the given procedures work.

Again, though, most likely you will be permitted to use calculators on your exams. Most exams are a test of concepts and not computation, and this book reflects as much. If you understand all the underlying math concepts you are being tested on, you will likely do well on your exams, and your calculator won't even be an issue. If you are going to use a calculator for your studies and on exams, be sure that you know how to use it correctly.

Unless you are doing advanced math, you are far better off with a very simple calculator than one with many special features which may be used incorrectly. For example, some calculators have a key that reduces fractions to lowest terms, but knowing how to correctly use that functionality can sometimes be more confusing and error-prone than just learning how to do it by hand.

Without a good grasp of the math itself, you will have no way of knowing if your calculator has given you an incorrect answer because you keyed in a problem wrong. Even if you are allowed to use a calculator, your goal should be to do most or all of each problem by hand, using your calculator to check yourself.

HOW TO USE THIS BOOK FOR SELF-STUDY

Unfortunately, many people will buy this book and simply not use it. Some won't look at it at all, but will momentary feel good that they have it on their shelf. Some will quickly flip through it and say, "Yeah, I know all this stuff," whereas all they really do is recognize the material or have a basic familiarity with it. Others may recognize what this book has to offer, but will simply not have time to read it since they are scheduled to take an important exam in just a few weeks if not a few days.

As obvious as this sounds, if you want to benefit from this book, you will have to read it. You will have to start at the beginning, and carefully read through it to the end. You will need to be disciplined about not skipping over any chapters or sections which you feel you already know. You will need to go through the material as slowly and as often as you need to until you have fully grasped all of the concepts, and can perform related problems quickly, easily, and accurately on your own.

This book only includes one or two sample problems for each topic for reasons described next. It is important that most of your time and effort be spent studying the actual concepts involved in each topic, instead of just copying the sample problems in your notebook over and over again. Doing that is an inefficient technique called "rote learning" which is discussed later in the book in the section on how to *not* study for math exams.

There are very few topics in math for which learning by rote is beneficial or even sufficient. As described in the next section, you really need to think about what you are doing, and what is going on with each problem. You also must constantly focus on the connections between all of the different topics so that you learn how they are related, and not get them mixed up.

THE BOOK'S POSITION ON PRACTICE EXERCISES

Despite what you may think, mastery of a topic usually does not come from doing repeated practice problems involving that topic. Rather than blindly doing 50 very similar practice exercises by rote, without any true understanding, it is better to do just one or two, and really think about what you are doing. The same principle is true of why you shouldn't obsess over the sample problems in any book as though the problems you will face on an exam will be identical. By adopting this mindset, you will be prepared for any variants of problems that you may face on an exam.

The best way to learn a topic is to just study the concepts that it is based upon, and then make up your own practice problems and try to solve them. Do your best to try to outguess the teacher or the exam you're preparing for. You can contact me for guidance, and it is also easy to do a simple Internet search to find a variety of practice problems on any topic. There is no need to utilize any pay service for such since so much free help is available.

The beauty of math is that all the procedures and formulas work no matter what particular numbers are involved. The goal is to master a topic to the point where you can apply it in any situation, and you won't panic if you encounter a similar problem with numbers that you haven't worked with before. It's somewhat like knowing how to drive your car, and knowing how to read a map and street signs, and therefore are prepared to drive wherever life happens to take you.

SCOPE OF OTHER BOOKS IN THIS SERIES

This book is the first in a series of at least three. The second book will cover the basics of algebra, and third book will cover the basics of geometry.

Most students should find that the information in the first three books is more than sufficient to fulfill the objectives of most math exams. It is worth noting that even graduate school entry exams such as the GMAT and GRE are only based on material that is approximately 10th grade level. However, those exams expect complete mastery of the concepts, and the ability to apply them in creative ways under extreme time pressure.

CHOOSING THE STARTING BOOK IN THE SERIES

No matter what your math level is, it is absolutely essential that you begin with this book. As described in the next chapter, the reason why math is hard for so many people is because they refuse to learn it from the beginning, and learn it thoroughly. Even if you are studying for a test in algebra or geometry, you should start with this book. Even if you took a placement test which states that you are at an advanced grade level in math, you should start with this book.

This is the path that you must take if you want to permanently and completely end the cycle of math failure once and for all. This series of books is a roadmap, and the road starts with this book.

HOW TO GET MORE HELP ON A TOPIC

I maintain a free math website of over 300 pages including the means for students to e-mail me their math questions. That will continue with the publication of this book, although I am working to redesign the website to better align it with the book. The old content will still be available, and new content will be added as students ask questions or make comments about the book and related material. My goal is for the website to serve as an interactive companion to the book so that students' questions can be addressed. The website and my question-answering service will continue to be free for all.

THE TARGET AUDIENCE FOR THIS BOOK

While I did not write this book with any particular audience in mind, it is best suited for older students, adults, and parents of students. It should prove useful to students who are studying for their end-of-grade (EOG) or high school exit exam, college entrance exam, (P)SAT, GED, or similar exams such as career-based ones.

The book should also prove valuable in a homeschooling environment of students of any age. The book will also likely be of benefit to teachers, especially new ones, including both math teachers who are looking for a different perspective, as well as teachers who are teaching math outside of their area of expertise due to a math teacher shortage situation.

Younger students are omitted from the suggested audience simply because they unlikely to sit down and read the book even though they could greatly benefit from doing so. The book does nothing to present math as something fun, exciting, or relevant to common everyday tasks. The reader simply requires the maturity to handle a book which effectively says, "Here is what you need to know, now go learn it."

Not only is this book appropriate for students who are studying higher level math such as algebra or geometry, it is absolutely essential that the material presented in this book be fully mastered before attempting to tackle those subjects. As mentioned, the main premise of this book is that it is a lack of basic knowledge which is at the root of the math struggle of virtually all students.

WHAT THIS BOOK WILL *NOT* ACCOMPLISH

The purpose of this book is to help you get through whatever math requirements you're facing. It won't make you a math whiz, and it won't result in very high grades if you are very far behind in math and have limited time in which to study. The book probably won't bring you to the point where you actually enjoy doing math. As the title suggests, after reading the book you will hopefully find that math has been made a bit easier, and hopefully it will help you achieve your math goals.

This book is certainly not intended for anyone who is pursuing a math-based degree or career. Such students need to study math from full-fledged textbooks as part of a highly-enriched classroom setting. Students who are pursuing math-based degrees or careers but require extensive math help should give some consideration to whether they have chosen the best path for themselves.

As discussed, this book will serve little to no benefit if you are scheduled to take an important exam tomorrow or next week, and are sorely unprepared for it. Flipping through the pages or sleeping with the book under your pillow is simply not going to help you. If you have bought this book as the result of such a "panic" situation, the limited time between now and your exam is best spent on mental/emotional preparation and learning test-taking techniques, as opposed to last minute academic cramming which will likely just fluster you even more. This is discussed more in Chapter Twelve.

THE BOOK'S POSITION ON LEARNING DISABILITIES

The subject of learning disabilities is extremely sensitive, and everyone has their own opinions on the matter. I firmly believe that while not everyone has the aptitude or interest to conquer college-level math, virtually everyone can learn to handle the math that is required to earn a high school diploma or similar. Some students must simply work harder than others to overcome whatever special challenges they may face. It is no different for math than for anything else in life. Every one of us has things that we find easy, and things that we find more challenging. We simply must work less on the former and more on the latter.

I also believe that a learning disability can sometimes become a self-fulfilling prophecy for some students. We are very quick to apply labels to things, and if the same labels are repeated often enough, we tend to integrate them into our sense of selves, and live up to them.

The path to learning math is exactly the same for every student, but each student will walk that path differently. Math must be learned thoroughly and step-by-step starting with the earliest material, and progressing systematically at whatever pace the individual student is capable of. Topics that the student finds easy can be covered quite quickly, and topics that the student is struggling with require additional time and effort, but certainly cannot be rushed or skipped over.

Rushing or skipping over topics sets the stage for struggling with math later on, at which time the math that was never learned in the first place must be learned. It is the proverbial "sweeping under the rug" which does not make the dirt actually go away.

The main reason why students struggle with math is because a typical school setting is not sufficient to teach math in the manner described above, and most parents cannot afford extensive and on-going private tutoring. This is discussed more in the next chapter.

THE BOOK'S POSITION ON EXPERIMENTAL CURRICULA AND "NEW" MATH

More and more schools are turning towards unconventional math curricula. This is yet another matter which is the subject of controversy and debate. Some programs make extensive use of tangible manipulatives which are intended to aid students who are considered to be visual learners. Other programs employ cooperative learning, group work, experiments, and student presentations. Still other programs attempt to teach math by way of elaborate projects which in some cases are integrated with coursework from other subjects and/or "real-world" scenarios. Some programs utilize obscure and contrived algorithms (procedures) for solving simple problems.

This book explains math concepts in a very straightforward, traditional, conventional manner. It is important to understand that a typical standardized test such as the SAT, GED, college entrance exam, or career-based exam does not provide any leeway for the use of any special physical props and/or special procedures. In fact it is often the case that using such devices is even more complicated, time consuming, and error-prone than just solving a problem in the traditional manner. The exams are simply designed to test whether or not the student can perform the given problems quickly and accurately.

As expected, the position of this book is that if you want to end the struggle with math, you simply must learn real math. I have personally observed countless students fail their exams while saying that the special devices and procedures used in class simply did not serve them on their exam. It is critical to avoid taking a basic topic such as adding fractions, and turning it into a huge production. While diagrams and tangible "fraction circles" are certainly beneficial as initial learning aids, at some point they must be set aside to transition to learning the basic procedure and all of its underlying concepts.

WHY IS MATH SO HARD?!

The next chapter attempts to answer the question of whether or not math is "hard," and what can be done about the matter in any case. It is important to not skip over that chapter as it continues to describe the entire philosophy upon which this book is based.

CHAPTER ONE

Is Math Hard, and If So, Why?

YES, MATH IS HARD

Most students describe math as being very difficult. For many, it is nothing short of torture. This book does not attempt to invalidate those feelings. However, it does attempt to explain them, and offer a solution for reducing and ultimately ending the suffering. It also outlines a roadmap which will hopefully lead students toward the successful completion of their math goals.

Most students struggle with math for a very specific reason, and that reason is the same for virtually all students. The reason and its causes are explained next.

THE ULTIMATE REASON WHY MATH IS HARD

Most students do not begin their education with the feeling that math is hard. Young children very quickly learn how to count to 10, and soon extend that limit to 20 and then 100. Almost intuitively, children begin to make connections between numerals and the quantities of objects that they represent. During this time the basic concepts of arithmetic begin to form naturally. Adding and "take-aways" make perfect sense, as do the concepts of arranging objects into equal groups. Math is not painful at this time, and if anything is fun.

For most students the first symptoms of struggling with math begin at about the third or fourth grade. Not only is more expected of students academically in general, but more material is crammed into less time. Students must now do a little bit of independent thinking and memorization. They must follow some basic multistep procedures without getting the steps or their order confused, and without confusing one procedure with another.

At this point the logistics and politics of education also come into play. Schools begin "teaching to the test" to avoid losing funding or incurring some other "punishment" which could even include legal issues if it is proven that not all of the material was covered. It is also during this time when the difference between a "good" and a "bad" teacher or school starts to become significant. While almost all math teachers know how to multiply 8 × 7, not all of them are capable of explaining what multiplication actually is. Even among the ones that can, many do not have a good sense of how that operation ties into to more advanced concepts that their students will be studying in later math.

At this stage most students start to develop confusion and misconceptions in math. Math is seen as an obscure and arbitrary field of study invented by the teacher and/or textbook publisher. The student quickly comes to view homework not as an essential stepping stone to later math, but simply as "busy work." Tests are seen as hurdles which must merely be "passed," perhaps with a boost from an "extra credit" problem, or a "make-up exam," or any other such grade manipulation devices.

Because of social promotion, students are "pushed through" from grade to grade, oblivious to the fact that they are falling deeper and deeper into a hole which they will ultimately have to dig themselves out of since no math is truly being learned. Certainly the student is "exposed" to various math topics throughout the years, and can remember having seen them at some point, but the topics are never actually mastered. It is as simple as that. Anything that hasn't been learned is hard.

HOW PARENTS CONTRIBUTE TO THE PROBLEM

A huge part of the problem is that most parents truly have no way of understanding that all this is happening. If Johnny brings home a 47 on a test, he may be scolded for not studying. His parents will then feel optimistic if he gets a 54 on the next test regardless of its topic or circumstances. If he gets a 71 on the test after that, his parents will probably celebrate and breathe easy. They will likely say that Johnny is obviously now focusing on his work, and is "more than passing." They may even suspend any tutoring services that they were paying for with the logic that they are no longer required.

This pattern continues throughout the years until at some point, usually sometime between fifth and ninth grade, Johnny starts bringing home grades between 0 and 40, and announces that he has absolutely no idea what is going on in class. At this time it is common for parents to seek out private tutoring, not for basic math skills, but for the particular topic that the the student is currently studying. For example, the parent may tell the tutor, "Johnny was always a good math student, but he's been getting low grades lately. He needs help with factoring polynomials since his teacher is giving his class a retest this Friday."

At this point a conscientious and independent tutor will try to explain the situation to the parent, but it often isn't well received. The concept of filling in gaps in basic math skills is not accepted. The parent often asks what sense it makes that the tutor wants to waste time and money on topics which the student not only was taught a very long time ago, but which have nothing to do with the important upcoming test that the student must pass in order to not have to retake the class in summer school.

Unfortunately this is a consistent pattern among many students and their parents. With this in mind, it is not hard to understand why so many students simply declare that "math is hard." The problem just "goes around in circles," and is never resolved.

As if the above points of concern weren't enough, the problem is further compounded by the fact that most of us want everything to be fast, fun, and easy. We want to use the minimum effort to achieve our goals.

GETTING DOWN TO THE ROOT CAUSE OF THE REASON WHY STUDENTS STRUGGLE WITH MATH

Most topics in high school math are based on topics which are assumed to have been learned and mastered in earlier grades. The word "mastering" means just that. It doesn't mean that the student merely remembers having been taught those topics, or that the student did learn those topics to some extent, but never fully "got them."

For any given topic, one must imagine all the prerequisite topics as forming a chain of links. If any link in the chain is broken or missing, the student will find the topic in question "hard" since s/he does not have the prior knowledge upon which the current topic is based. Unfortunately, most teachers and textbooks do not emphasize the links to later topics at the time that a topic is taught, nor do they emphasize the "back-links" for the current topic which must solidly be in place.

The field of math is truly a giant matrix of interconnected links, but instead is often taught in the same manner as one might teach vocabulary words for a foreign language class. The lesson on learning the names of colors has no connection between the lesson on learning the names of animals. Math simply doesn't work that way, but far too few people understand and accept that.

What all this means is that for math to stop being "hard," all of the links have to be repaired and/or firmly set into place. For some links this will be fast and easy. Sometimes it only takes five minutes for a tutor to clear up a student's misconception about a topic. Sometimes it can take dozens of sessions for just that one topic, especially if it is something very broad, or if it was never learned at all in the first place.

Sometimes a topic needs to be traced all the way back to the very first links in the chain, and in doing so it is determined that the student has a very poor command of material that is even eight grades below the grade level of the math that s/he is presently studying.

If you want for math to stop being "hard," and if you want to be able to approach your exams with confidence as opposed to the feeling of "rolling the dice" or praying for divine intervention, you will simply have to accept the "chain link" model of how math is learned, and seek out help accordingly.

This means that it will likely be the wrong approach for you to buy an algebra review book because you are "stuck" on algebra, or to seek out an algebra tutor to help you with your current algebra topic. Instead you must change your mindset to seeking out help in repairing and filling in the gaps in your math knowledge, no matter how long it takes, and no matter what effort is required on your part.

SO IS MATH *REALLY* THAT HARD?

As dismal as the previous sections may have come across, the reality is that there really is not that much math material required to earn a high school diploma or pass an exam that tests the equivalent knowledge. The average school period is about 45 minutes, and each year there are only about 120 class days spent on actual new math lessons. The remaining days are taken up by review sessions, tests, and other activities.

It is worth noting that many new lessons are nothing more than a slight extension of a previous lesson so they aren't completely new. Even for the lessons that do present totally new material, the material presented in most lessons is simply not that lengthy nor that "hard." A lesson often revolves around just one small definition, formula, or procedure. Topics that are more involved are spread out over many lesson days. Also, remember that if you are going to study on your own or with the help of a tutor, you can obviously go through the material much faster than can be done in a typical school class.

What it all comes down to is that even if you are eight grades below level in math, it certainly does not mean that it will take eight calendar years to catch up. Conversely, though, it doesn't mean that you can catch up in just a few weeks, and perhaps not even a few months either. It will not take "years and years," though, as some students unnecessarily fear.

College-level math can certainly be quite abstract and challenging, and quite frankly, is just not for everyone which is why not everyone chooses to pursue a math-based career. However, the math that is required to earn a high school diploma or pass exams at a similar level is not at all as insurmountable as it may seem. The premise of this book is that fulfilling your current math goal is very much "doable," and it doesn't have to be painful.

Again, the most important thing is to just get yourself into the mindset of repairing and filling in gaps. A good way to start is by reading this book cover to cover, and not just turning directly to the chapter that you feel is best to start with. Start at the first topic and assess yourself honestly to see if you have truly internalized the material being presented. If you have, then move onto the next topic, and all that happened is you spent some valuable time reviewing and reinforcing an earlier topic. If a topic is not fully clear to you, spend as much time on it as necessary until you have fully mastered it.

WHAT IF A STUDENT IS ONLY SLIGHTLY BEHIND?

In all of my experience tutoring, I have yet to work with a student who is just "slightly behind" in math. On the contrary, virtually all of my students were/are below their chronological grade level to the extent of anywhere between one and as much as eight grades. This is why students find math hard. If a student is very much below grade level, but insists on seeking out math help in the same manner as a student who is truly only "slightly behind," that student will not obtain the proper help that s/he needs, and will continue to struggle with math.

ASSESSING YOUR OR YOUR CHILD'S MATH LEVEL

Unfortunately there is a great deal of politics involved with math education. While your child's school or teacher may state that your child is on a particular grade level, that grade level will likely be different than what is stated by the county, state, and federal government, all of which may not even coincide with one another.

The purpose of this book is not political. The purpose is to help make math a bit easier for you, and to explain math topics in plain English. With that in mind, by far the best thing to do is to dismiss all notions about your or your child's supposed grade level in math. It simply doesn't matter except to those who have a vested interested in the politics of education. It also doesn't matter what grade level is associated with a given topic. You have either mastered a topic or you haven't. If you haven't, you'll have trouble with later work so you must do so now, and not push it aside or dismiss it.

As I describe later in this chapter, this book lays out a roadmap for you or your child to follow. If you can forget about test scores, and grade levels, and any other contrived statistical information that has no relevance, you can put all of your effort into learning the math that you probably never fully learned in the first place. In this manner, math will make more and more sense, and will become less and less frustrating .

ISSUES WITH MANIPULATION OF TEST SCORES

It is common for parents or adult students to discuss exam scores as though they hold some sort of divine, ultimate truth. Of course this serves no purpose whatsoever, and in no way addresses the issue of what to do about their or their child's struggle with math.
It is worth noting that test scores are extremely easily manipulated in many ways and on many different levels including teacher, school, district, state, and federal.

Unfortunately this creates tremendous confusion for students and parents. The matter could easily be the subject of its own book, but elaborating serves no purpose. Again, to end the struggle with math, the focus simply must change from exams and exam scores to the actual math itself. Leave the politics to the politicians.

A PASSING MATH GRADE CANNOT BE 65

This section admittedly invalidates the previous one to some extent, but some important points must be made. By far the biggest problem with our grading system is that we usually consider 65 to be a passing grade for all subjects, at all levels, regardless of context. The following example is very important to fully understand. If you are a computer science major in college, but are required to take a liberal arts class in music or art, there is not much harm in earning a 65 or "C" in that class. Certainly it will pull down your average a bit, but you can still graduate with the highest honors, and the material in that class does not relate to your chosen career path or field of study. The music or art class is just a single, isolated class, intended for academic enrichment.

That is a very different scenario than the math classes which students take throughout their schooling. It is easy to understand that if a student has only mastered 65% of second grade work, that student will enter the third grade at a significant disadvantage. While some of the previous grade material will be reviewed, it is often done so quickly, and in the context of introducing new material. The process of falling behind begins.

If that student manages to pass third grade math with a 65, obviously the problem continues into fourth grade, and it keeps compounding and compounding. By the time the student has been "pushed through" the system into middle or high school, his/her math education is full of gaps, and s/he must now fight to overcome and fill in those gaps while attempting to learn abstract math in the context of algebra and other higher level topics.

Since math continually builds upon itself from one grade to the next, it is simply not OK to be promoted from one grade level the next without understanding virtually all of the material. Certainly there needs to be some leeway for a student to make careless mistakes, and to not fully master every single topic, but for the most part, the math at each grade level truly must be mastered and internalized in order to tackle higher level material which builds directly upon lower level material.

COMING TO TERMS WITH YOUR OWN MATH LEVEL

When students first reach out for formal help with math, they often have lots of difficulty coming to terms with their math level if assessed in an unbiased manner. Parents usually have an even harder time with this.

From my experience, the actual math level of the average 9th grader is somewhere between 3rd and 7th grade. I define this level as the student's ability when tested under true, formal test circumstances. It is important to understand that even if very slight assistance is provided during an exam, the entire result can become very much skewed. Such assistance can even come in the form of something seemingly innocuous such as the teacher's intonation when reading a question to the student, or the teacher saying, "Johnny, don't you remember, we did a problem exactly like this on Monday."

Of course what happens is that at some point the student must take a standardized test such as the SAT or a career-based exam which is administered in a highly controlled environment, and then the student quickly realizes that s/he isn't able to answer the questions on his/her own. While there are many ways that the student can rationalize doing poorly on such an exam, it doesn't change the fact that s/he didn't attain his/her goal.

While it serves no purpose to embarrass a student by implying that s/he is six grades below level, if a student is to ever receive any help, and end his/her struggle with math, a proper starting point must be established.

For example, if a student informs his/her tutor that s/he is having trouble dividing algebraic fractions, but the student doesn't know how to divide even simple numeric fractions, the student must be informed of and accepting of that fact, and that will become the starting point. If it is determined that the root of the problem is even further back such as the student having virtually no concept of division at all, let alone division involving fractions, then that must be the starting point. The starting point must be as far back in the hierarchy of topics and levels as necessary.

All of this can be a great attack on one's ego, especially in cases where the student and his/her parents have spent years under the impression that everything was fine. It is not uncommon for a parent to inform a tutor that their 11th grader has taken and passed honors math classes since elementary school, yet the student is incapable of adding two negative numbers together.

The response to the situation is quite parallel to the emotions that people typically experience in response to receiving bad news on a much grander scale. The initial response is denial. That is usually followed by anger, first at the individual who performed the student's assessment, and perhaps later at the education system in general. There is then often some bargaining which comes in the form of the parent saying something like, "Maybe if we just work on the material for Johnny's final exam which he has to take in two weeks, he'll manage to pass it, and then next year he'll have a different teacher and everything will be fine." At some point feelings of helplessness and hopelessness and depression set in.

Eventually, though, if the student is ever to get the help that s/he needs, there will need to be a sense of acceptance on the part of the student and his/her parents. It is just as simple as that. Until that happens, the student and his/her parents will be wasting their time and money in seeking out any special extra help.

RECOGNIZING OR BEING FAMILIAR WITH A CONCEPT ISN'T THE SAME AS KNOWING IT

To help drive home the important premise of this chapter, it is worth restating that recognizing or being familiar with a topic is not the same as knowing it, but the two are often taken to be equivalent.

For example, a student may tell his/her tutor that s/he has trouble computing the area of a circle, but then chastise the tutor for explaining pi (π). The student may say something like, "We already did pi in class. Three point one four. Just teach me the formula since it will be on tomorrow's test."

This is an example of recognizing a concept, but not actually understanding it. Johnny remembers having seen the π symbol on the board, and remembers his teacher talking about it, but s/he has no idea of what it is, or how or when to use it. This will almost always lead to the student getting related exam questions wrong which is then rationalized by saying, "The questions on the test weren't like the sample problems that were in the book." It simply means that the student never truly learned the involved concept, but only was able to recognize it.

HOW EXAMS CONTRIBUTE TO THE PROBLEM

Most school tests focus on topics in a "vacuum." For example, a test may be given on how to find the volume of a cube, or how to divide a fraction by a fraction when negative numbers are involved. It is common for the previous day's lesson to be spent reviewing the topic, in some cases working from a practice test which is virtually identical to the test that will be given the next day.

Of course what happens is that the topic ends up being learned by rote, and is "spit back" on the test the following day. If one or two "trick" questions involve some unexpected numbers or some other twist, those questions are typically gotten wrong, but the student still ends up passing which is all s/he was concerned about. The test is over with, and the topic is quickly forgotten to make room for new topics to be handled in this manner.

At some point the student is asked to prepare for a comprehensive midterm or final exam. While a review may be conducted in class the previous days, the homework assignment for those nights will simply be "Study for test." The student will typically go home, flip through his/her textbook and notebook while saying, "Yeah, I know all this stuff." As described, of course, almost always it is not actual knowing, but simply recognizing. If the student's parents question him/her because they don't see any homework problems being worked on, the student will simply retort that there was no homework.

The next day the student takes the test, and rationalizes a low grade by saying that s/he forgot the earlier material because it was taught so long ago, and because the questions given in a different order than the topics were taught, and so s/he got confused. It is then common for the teacher to give the class a "retest," or allow the students to reclaim lost points by submitting "test corrections" or some type of extra credit assignment.

After all of the logistics are finalized, the end result is typically that the student being passed on to the next grade. Little to no math was actually learned, yet the student must now take on more challenging math which is completely dependent upon having learned the prerequisite material in the previous grade. The student of course will simply say that the new math is "hard."

THE "CATCH-22" OF TEACHING AND TUTORING

There is an inherent flaw in most models of teaching and tutoring which contributes to the reason why students do not learn math properly. It is also the reason why students don't get the help they need once they reach the point where they declare math to be hard.

The problem is that we expect teachers and tutors to function in the capacity of "helpers." No teacher, independent tutor, or commercial tutoring center wants to be accused of, or earn a reputation as "not helpful" or "refusing to help."

The end result is that those who are expected to help tend to take the path of least resistance which is often what the student and his/her parents actually want and expect. The student says, "I need help with factoring polynomials," and the educator pulls out a worksheet on the specific topic—not on the prerequisite topics which usually must be addressed and mastered first.

One problem at a time, the teacher or tutor handholds the student through the solving process. Any stumbling blocks are quickly glossed over by the instructor, either by providing a detailed hint or by outlining the entire thought process that is required. In extreme cases the actual answer is just provided outright, and the student is deluded to believe that s/he did all the work.

In any case, the problems are "worked through" by the student and teacher or tutor, leaving the student with the impression that s/he "gets it now." If the teacher or tutor elicits any information at all from the student, it is minimal. Most of us want to be given the proverbial "fish for a day." We don't want to be taught how to fish for ourselves, and we certainly don't want to be taught the thought process behind learning how to fish.

A typical session of private tutoring or after-school help ends with a smile on the face of the student along with a feeling of confidence and accomplishment. The student may proudly inform his/her parents that s/he "gets it now." Everyone walks away happy. Unfortunately, if the session followed the pattern described above, nothing was truly learned. All that took place was the illusion of learning. The student then goes on to fail his/her next exam, and all those concerned are baffled.

The point to take away from all this is that for true learning to take place, a student must be forced to think for him/herself, as unpopular as that concept may be. The student must learn how to solve his/her own problems, with legitimate guidance along the way, as opposed to just being robbed of any opportunity to think and wrestle with a problem. It is similar to the technique that therapists employ when they respond to a patient by asking, "Why do *you* believe that you feel that way?" As mentioned, though, very few teachers, tutors, or commercial tutoring services are willing to take the risk of utilizing such a model of instruction.

THERE IS A SOLUTION TO THE PROBLEM OF STRUGGLING WITH MATH

Struggling with math is not comparable to dying of an incurable disease. While the latter may have to be accepted, the former does not. The essential first step to ending the struggle is to accept that you are simply going to have to learn all of the math that you never fully learned. You will continue to struggle as long as you convince yourself that you "know" concepts that you don't truly know, and as long you endlessly search for the miracle tutor or miracle textbook that will provide instruction on whatever topics you are currently responsible for knowing. You will have to accept the seemingly paradoxical notion that to master your current topics, you will have to set them aside and focus on the earlier topics that form the basis of your current ones.

THE ROADMAP FOR MATH SUCCESS

As mentioned, this book was designed to be roadmap. Begin by thoroughly reading and studying the next chapter which covers basic skills in arithmetic. Progress through the remaining chapters as slowly as you need to, making sure that you absorb everything along the way.

The final chapter provides some suggestions about how to learn math and study for exams, and you can certainly read that chapter out of order from the others. The book concludes with information about the book's companion website including how to contact me with questions about the material or if you would like to check your understanding.

CHAPTER TWO

The Foundation of Math: Basic Skills in Arithmetic

WHAT ARE NUMERALS?

We see and use numerals every day, but we usually don't think about what they really are. A **numeral** is just one or more symbols that we use to represent a particular quantity of things. For example, we use the symbol "3" to represent this quantity of happy faces: ☺ ☺ ☺

Our number system uses ten symbols or **digits** (0, 1, 2, 3, 4, 5, 6, 7, 8, 9) which we combine to represent different quantities. To represent a quantity greater than nine such as ten, we must combine a 1 and a 0 each in its own place since we don't have an individual symbol for ten.

Interestingly, our math system is based on groupings of 10. Ten groups of 1 is 10, ten groups of 10 is 100, ten groups of 100 is 1000, and so on. This forms the basis of our **place value** system which we'll work with later.

THE (POSITIVE) NUMBER LINE

As you do basic arithmetic, it's important to have a mental picture of a standard number line. It's not practical to actually count on one, but know that we can represent numbers by writing a 0 on a line, and then marking off higher and higher numbers as we move to the right. Later we'll talk about how we can represent negative numbers to the left of zero, essentially forming the mirror image of what is on the right.

ADDING SINGLE-DIGIT NUMBERS

Adding is just a fancy word for combining. If you have 2 objects, and someone gives you 4 more, you'll need to **add** them to get your new total. We can use a number line to add, although it's rather cumbersome. Since we started with 2 objects, start at number 2 on the line. To add 4 more, move four tick marks to the right which us to our answer of 6.

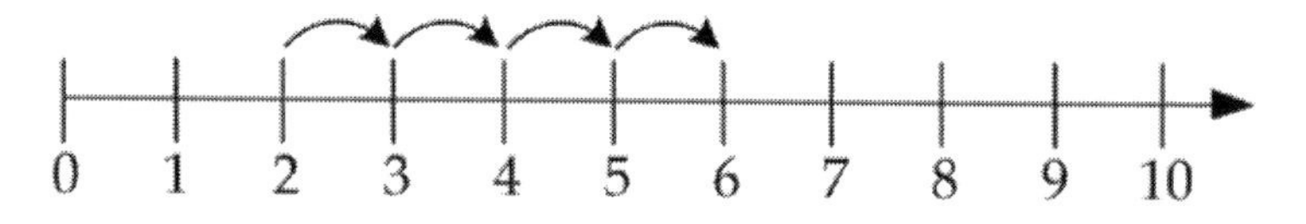

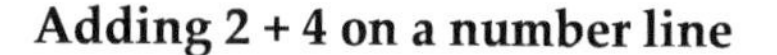

Adding 2 + 4 on a number line

Instead of using a number line, we could simply use our fingers to add small numbers, or count in our heads which would be even faster. We start at 2, then count four numbers: 3, 4, 5, 6. We end up on 6 which we say is the **sum** of 2 plus 4.

We don't want to always have to add by counting on our fingers or in even in our heads since that takes time and can lead to errors. It's best if you can memorize some basic addition facts so that you won't always have to compute them every time they come up. The way to get better at that is by practicing.

It's helpful to use index cards to make flashcards for basic addition facts with the problems on the front and the answers on the back. Use them to quiz yourself on every combination that involves adding a number between 0 and 9 to another number between 0 and 9. Some examples are 8 + 5, 4 + 7, 9 + 9, 6 + 0, etc.

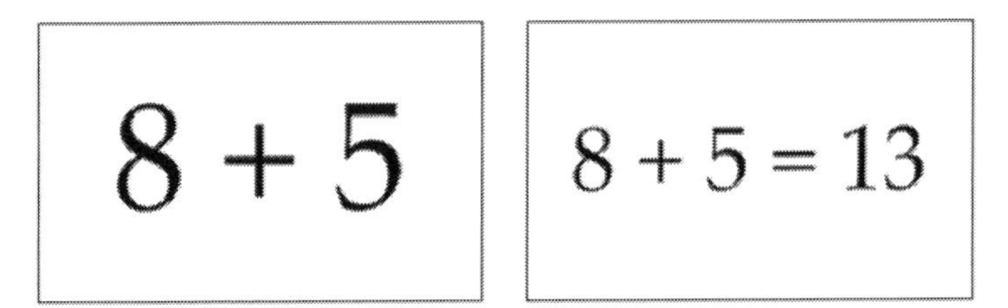

Use the front and back of index cards to make flashcards

There are some tips that can make this task much easier. First, you probably feel comfortable with the fact that 2 + 5 equals 5 + 2. We're just combining a quantity of 2 with a quantity of 5. The order that we do it in doesn't matter.

Later we'll use the word **commutative** to describe this property of addition. It just means that the order doesn't matter. Once you know an addition fact one way, you automatically know it the other. Not all operations are commutative which we'll see later.

You probably know that if we add 0 to a number, it doesn't change it at all. We added nothing, so we're back to where we started. 9 + 0 still equals 9.

There is a useful shortcut you can use when you have to add 9 to a number. Think about adding 9 + 7. It's inefficient to have to count up 7 numbers starting at 9.

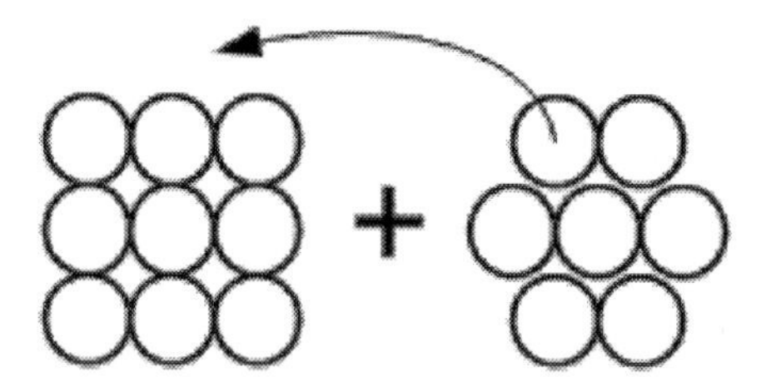

Adding 9 + 7 is the same as adding 10 + 6

Think about how we can regroup the problem. We could take one of the 7 objects we're adding, and move it into the group of 9. That would reduce the 7 objects to 6, but increase the 9 objects to 10, making an equivalent problem of 10 + 6. We haven't actually changed the problem. We just rearranged it to make it easier to get the answer.

SUBTRACTING SINGLE-DIGIT NUMBERS

Subtracting just means taking away. You start with a certain number of items, and then some or all of them are removed. We saw that to add, we count by moving to the right on a number line. To **subtract**, we do the opposite—we move to the left. As simple as this sounds, it is very important to keep this in mind later when we start to subtract numbers which will lead to special situations.

Let's try doing the subtraction problem 7 – 3 on a number line. We start at 7, and then move three tick marks to the left, ending up on 4 which is our answer. Of course we could have used our fingers by starting with seven of them raised, and then lowering three. Obviously both methods are cumbersome and impractical.

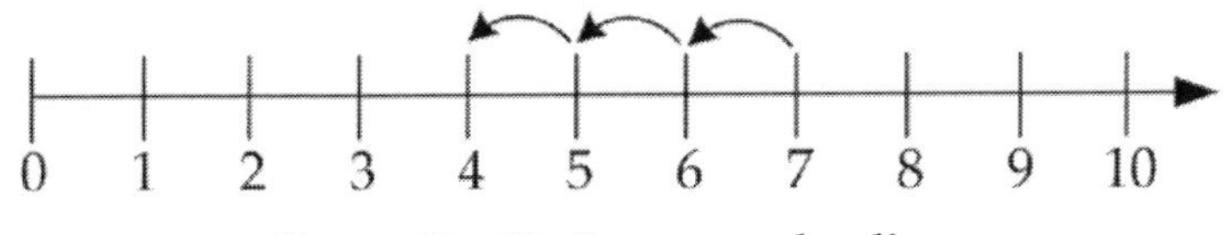

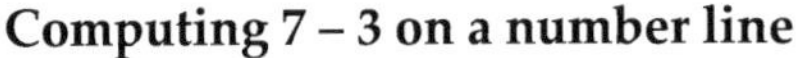

Computing 7 – 3 on a number line

It's best if you can memorize some basic subtraction facts so you won't have to recompute them every time. The way to get better at that is by practicing. As with addition, it will be helpful to use index cards to make flashcards with basic subtraction facts. Quiz yourself on every combination that involves starting with a number between 0 and 9, and subtracting a number between 0 and 9 that is *less than or equal to* the first number. Some examples are 8 – 5, 7 – 4, 9 – 9, 6 – 0, etc.

Later we'll learn how to deal with situations where you must subtract a larger number from a smaller number, such as 2 – 5. We cannot take it upon ourselves to reverse the problem and make it 5 – 2. That is completely wrong. It is also incorrect to say that the problem of 2 – 5 is "unsolvable," or that "you're not allowed to do it." We'll discuss this in detail in Chapter Four.

We know that addition is commutative—the order in which added two numbers doesn't matter. Subtraction is *not* commutative. The answer to 7 – 4 is not the same as the answer to 4 – 7, yet the latter problem is completely legitimate, and will be discussed later. We'll also discuss the fact that we refer to the answer in subtraction as the **difference** which can lead to some confusion later on.

ADDITION AND SUBTRACTION ARE OPPOSITES

It's important to understand that addition and subtraction are opposites. For example, we can add 2 + 7 to get 9. Now that we have 9, if we want to get back to 2, we have to subtract 7. We must *undo* the addition of 7 which was done to the 2. We say that addition and subtraction are **inverse operations**. This means is that they "undo" each other.

When you work with your subtraction flashcards, think about the related addition problem for each one. For example, to compute 8 – 6, it is helpful to ask yourself, "What number must I add to 6 in order to get back to 8?" The answer is 2, which is the answer to the subtraction problem. We just looked at it in reverse.

If we subtract 0 from a number, it doesn't change it at all. You took nothing away, so you're back to where you started. 9 – 0 still equals 9. However, 0 – 9 is a completely different problem which will be discussed later.

Note that the "regrouping" shortcut which we used for an addition problem like 9 + 7 does **not** work for subtraction. We're not combining so if we changed a problem like 9 – 7 into 10 – 6, it would give us the wrong answer.

ADDITION AND SUBTRACTION FACT FAMILIES

Recall that subtraction is the inverse of addition. When subtracting, you can approach the problem in reverse, and determine what number you must add to get back to the starting number. These facts come together to form a **fact family** for a set of numbers. See the example below.

$3+4=7$	$4+3=7$
$7-4=3$	$7-3=4$

An addition/subtraction "fact family"

MULTIPLYING SINGLE-DIGIT NUMBERS

It's important to understand that the operation of **multiplication** is nothing more than repeated addition. It's just a shortcut so we don't have to repeatedly add the same number to itself in situations where it's called for.

A common example is where there are 5 children, and you want to give 7 pieces of candy to each child. How many pieces of candy will you need? Of course you could just compute 7 + 7 + 7 + 7 + 7 to get your answer. However, if you know your multiplication facts, you can have the answer instantly.

Once you know that the above answer is 35, you never need to compute it again. You now know that 5 groups of 7 make 35. We say that 7 **times** 5 equals 35. For now that will be written as 7×5, but in later math we will instead use other symbols since the letter *x* will take on a special new meaning.

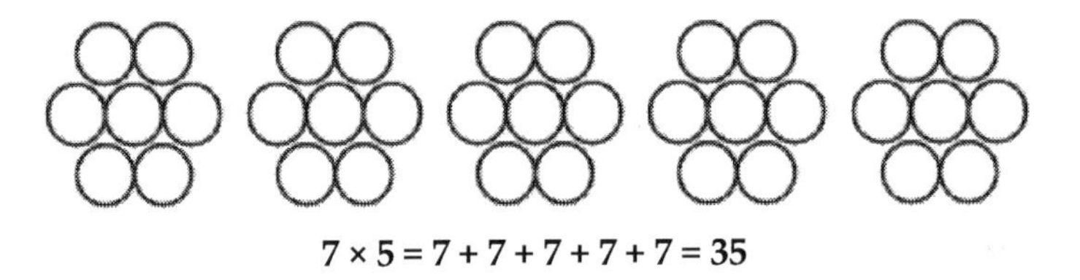

$7 \times 5 = 7 + 7 + 7 + 7 + 7 = 35$

We can do the above calculation for other combinations of digits. For example, to compute 6 groups of 9, we can add $9 + 9 + 9 + 9 + 9 + 9$, giving us 54. We now know that 9×6 equals 54, and we never have to compute it again.

There are some shortcuts to simplify the task of learning multiplication facts. Any number times 1 is simply itself. For example, if you have 1 group of 8, you just have 8.

Since addition is commutative, it follows that multiplication would also be since it is just repeated addition. This means that $4 \times 3 = 3 \times 4$. Three groups of four is the same as four groups of three. Stated another way, $(4 + 4 + 4) = (3 + 3 + 3 + 3)$. Once you learn a multiplication fact, you automatically know it in reverse.

MAKING AND USING A MULTIPLICATION TABLE

It's important to learn how to create a homemade multiplication table, and study it until it is memorized. Look at the table below. To multiply two numbers, find the first number along the row headers, and the second number along the column headers. Find the cell where the row and column intersect. That cell contains the answer to the multiplication, known as the **product**.

×	**1**	**2**	**3**	**4**	**5**	**6**	**7**	**8**	**9**	**10**	**11**	**12**
1	1	2	3	4	5	6	7	8	9	10	11	12
2	2	4	6	8	10	12	14	16	18	20	22	24
3	3	6	9	12	15	18	21	24	27	30	33	36
4	4	8	12	16	20	24	28	32	36	40	44	48
5	5	10	15	20	25	30	35	40	45	50	55	60
6	6	12	18	24	30	36	42	48	45	60	66	72
7	7	14	21	28	35	42	49	56	63	70	77	84
8	8	16	24	32	40	48	56	64	72	80	88	96
9	9	18	27	36	45	54	63	72	81	90	99	108
10	10	20	30	40	50	60	70	80	90	100	110	120
11	11	22	33	44	55	66	77	88	99	110	121	132
12	12	24	36	48	60	72	84	96	108	120	132	144

A 12 × 12 Multiplication Table

It is good practice to make your own multiplication table. Draw a 13 by 13 grid of blank cells modeled after the one above. Put a times sign (×) in the upper left hand corner, and number the column and row headers as shown.

We could easily make a table with more rows and columns to accommodate larger numbers, but multiplication tables typically limit themselves to products up through 12 × 12. This may have something to do with the common usage of the dozen (12).

We can easily fill in the values across the 1 row. 1 times any number is equal to that number. Since multiplication is commutative, we can also fill in the values down the 1 column. Any number times 1 is equal to that number.

Now let's fill in the 5 row. We already have 5 × 1 filled in. It's easy to compute a group of 5 repeated twice. That equals 10, so we can write that in the cell at the intersection of the 5 row and the 2 column. If you continue, you'll see that 5 repeated 3 times is 15, 5 repeated 4 times is 20, and so on. Notice that as we move along the 5 row, we're essentially just counting by fives. That makes sense since we're just increasing the amount of groups of five.

Remember again that since multiplication is commutative, whenever we fill in a row of the table, we can also automatically fill in the corresponding column.

LISTING THE MULTIPLES OF A NUMBER

As we move across a row or down a column in the table, we see the **multiples** of the number in that row or column header. For example, (7 × 1) is 7, (7 × 2) is 14, and (7 × 3) is 21. The numbers 7, 14, and 21 are all multiples of 7. We could also say that if we count 7, 14, 21, 28, and so on, we're "counting by sevens." We keep adding 7 more to the previous total. We continue counting by sevens to fill in the rest of the 7 row and 7 column.

We can continue to fill in the multiplication table in this way. Counting by twos is quite easy. We're not used to counting by threes and fours, so give that some extra practice as you fill in the related cells.

Counting by tens is easy. Elevens follow a pattern until you pass 99. As you count by nines, you'll notice an interesting pattern such that the digits in each product always add up to 9. For example, 9 × 8 = 72, and 7 + 2 = 9. Continue working in this way until the entire table is filled in. Make it a point to practice counting by different numbers by reading across their rows or down their columns, eventually being able to do this from memory. This skill will be highly important in later math, and will be used very frequently.

It is important that you be able to do this quickly and easily for numbers between 1 and 12, and that you can instantly recognize whether or not a particular number is a multiple of a given number. For example, you should reach the point where you immediately know that 48 is a multiple 6 and 8 (as well as a multiple of many other numbers), but is not a multiple of 7 or 9 since it cannot be found in the rows or columns for those numbers.

The multiplication table needs to be memorized to be able to effectively progress to later work. It should not be thought of as "busy work," but as a stepping stone.

LEARNING THE MULTIPLICATION FACTS

As with addition and subtraction, it's important to learn and memorize the basic multiplication facts. Make index cards as previously described, and study problems involving multiplying numbers between 0 and 12 by numbers between 0 and 12.

There are some tips that will make this job much easier. First, remember again that multiplication is commutative. The order in which you multiply two numbers doesn't matter. Once you have memorized the answer to 3×6, you automatically know the answer to 6×3.

Remember that we can always find the product of two numbers by counting through the multiples of one of them. For example, to multiply 4 × 6, start with the larger number, 6, and then count off the first 4 multiples. We would count 6, 12, 18, 24, to get to the answer of 24.

Some multiplication facts are best to just memorize. An example would be 8 × 7. It isn't practical to count off seven multiples of 8 or eight multiples of 7. It also isn't any easier to remember the product of 7 × 8 than it is to remember 8 × 7. It's just easier to memorize that the answer is 56.

Any number times 0 equals 0. Remember that multiplication is the process of adding up repeated groups of a quantity. What do we get if we try to add up 0 groups of 23? The answer is 0. We have nothing at all.

This is different than how 0 works in addition and subtraction. If you add 0 to a number, or subtract 0 from it, you get the number that you started with. This is summarized below using n to represent any number.

$$n + 0 = n \qquad n - 0 = n \qquad n \times 0 = 0$$

The effects of adding, subtracting, or multiplying by 0

DIVISION IS THE INVERSE OF MULTIPLICATION

Just like multiplication is repeated addition, **division** is actually just repeated subtraction, although in practice there are other, perhaps better ways of thinking about it.

Imagine you have 30 apples, and you want to divide them evenly between yourself and your friend. How many apples will each of you will get? One method is to repeatedly subtract 2 from 30, since we're trying to divide 30 between 2 people. Each time you subtract 2 apples, you can split them evenly between yourself and your friend as in "one for you, one for me."

If you keep a running tally, you'll see that you can subtract 2 a total of 15 times, and then you'll be out of apples. You'll each end up with 15. That means that 15 is the answer to the problem of 30 divided by 2, represented as $30 \div 2 = 15$.

Even though division is nothing more than repeated subtraction, in practice it's better to think of division as the inverse of multiplication, just like we saw that subtraction is the inverse of addition. Recall that to compute $7 - 3$, we could ask ourselves what number we must add to 3 in order to get 7.

Division and multiplication have that same relationship. They are inverse operations. For example, if we want to compute $24 \div 3$, rather than seeing how many times 3 can be subtracted from 24, we can determine what number must be multiplied by 3 in order to get 24. It is 8, which is the answer to our division problem.

TWO WAYS OF REPRESENTING DIVISION

Make sure you understand the connection between the division notation that we've been using and the notation that is taught in the lower grades. In the samples below, a is the **dividend** (what we are dividing), b is the **divisor** (what we are dividing by), and c is the **quotient** (the solution to the division problem).

$$a \div b = c \qquad b\overline{)a}^{\,c}$$

Two equivalent ways of representing division

DIVISION IS NOT COMMUTATIVE

Recall that addition and multiplication are commutative, but subtraction is not. For example, $3 - 5$ does not equal ($\neq$) $5 - 3$. The order in which we subtract numbers does matter which will be discussed in Chapter Four.

$$3 - 5 \neq 5 - 3$$

Like subtraction, division is *not* commutative. The order in which we do our division does matter. $12 \div 3$ does not yield the same answer as $3 \div 12$. The first example has an answer of 4, and the second one has an answer of ¼. That will be explained in Chapter Five.

$$12 \div 3 \neq 3 \div 12$$

MULTIPLICATION AND DIVISION FACT FAMILIES

We know that multiplication is commutative. We can swap the order of the involved numbers. We also know that division is the inverse of multiplication. When dividing, we can approach the problem in reverse, and determine what number we must multiply the second number by in order to get back to the first number. Those facts come together to form a "fact family" for a given set of numbers. An example is shown in the following chart.

$3 \times 6 = 18$	$6 \times 3 = 18$
$18 \div 6 = 3$	$18 \div 3 = 6$

A multiplication/division "fact family"

DIVISION WITH A REMAINDER

Quantities don't always divide evenly. A simple example would be if you wanted to evenly divide 3 apples among

2 people. Each person would get one apple, and then to be fair, the remaining apple would have to not be given to either person, or it would have to be split in half with each person getting one of the halves.

If you know your multiplication facts, you'll be able to determine if a division problem will have a remainder. Remember that division and multiplication are inverse operations. If a number can be divided evenly, there will be some way of reversing the problem into a multiplication problem involving whole numbers (not fractions), as described above in the section on fact families.

Think about the problem $13 \div 3$. We're trying to evenly divide 13 apples among 3 people. There are two ways of thinking about it. One way is to think of division as repeated subtraction. How many times can we take away 3 apples from 13 apples, yet still be left with apples to take away? The answer is 4. After taking away 3 apples 4 times we will have taken away 12, with 1 left over. For now, we can say that the answer to the problem is "4 remainder 1," abbreviated 4 R 1.

Recall that it's best to think about division as the inverse of multiplication. The question is, "What number can I multiply 3 by so that I can get as close to 13 as possible, but without actually going over it?" The answer again is 4. That gets us back to 12, but we have 1 left over. The 3 people will each get 4 apples, with 1 apple unaccounted for. For now we can call it a remainder of 1 and leave it at that, but in Chapter Five we'll learn how to convert that remainder into a fraction to see how the last apple can be split up and shared evenly.

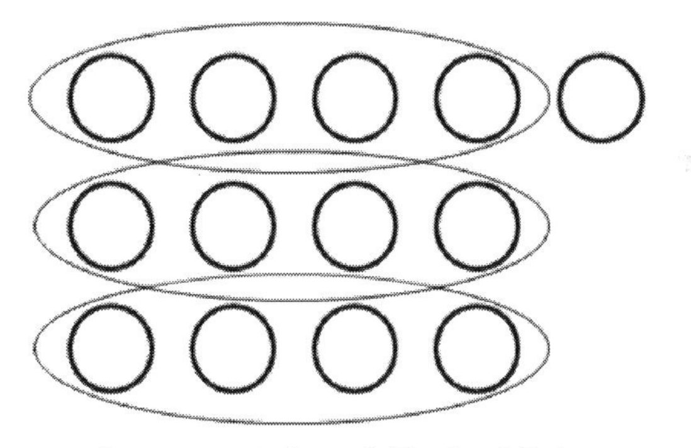

A representation of 13 ÷ 3 = 4 R 1

PLACE VALUE TO 1000

Our math system is based on groupings of 10. We start counting with single-digit numbers in the **ones** or **units** place. Think of it as a cash register compartment for the $1 bills. When we reach 10, we have to start a new column to the *left* of the ones place known as the **tens** place. Think of it as the compartment for the $10 bills.

Once we reach ten groups of 10, we have to start yet another column to the left of the tens place called the **hundreds** place (the compartment for the $100 bills). In Chapter Three we'll learn about how the pattern continues to thousands, ten thousands, and so on. Remember that each column can only contain a single digit.

In terms of a cash register, imagine that each compartment can't hold more than 9 bills. Once a compartment has 10 or more of the same denomination of bill, we must trade in groups of ten for bills of a larger denomination, starting or adding to a compartment on the left. Keep in mind that we only work with denominations of $1, $10, $100, $1000, and so on.

The following chart represents the number 3456 (three thousand, four hundred fifty-six) which is literally 3000 + 400 + 50 + 6 when examined in terms of **place value**. Each column is worth 10 times the column on its *right*.

Thou-sands	**Hundreds**	**Tens**	**Ones/ Units**
3	4	5	6
3000	400	50	6

The number 3456 broken down by place value

With money we have denominations such as $5, $20, and $50, but in math we only work with 1, 10, 100, 1000, and so on. All numbers must be represented in those terms.

DOING BASIC ARITHMETIC BY HAND

As explained in the Introduction, most likely you will either be permitted to use a calculator on your exams, or you simply won't need one. However, there is instructional benefit from learning how to do certain basic arithmetic problems by hand. This serves to reinforce many of the concepts presented so far.

TWO-DIGIT ADDITION WITH CARRYING

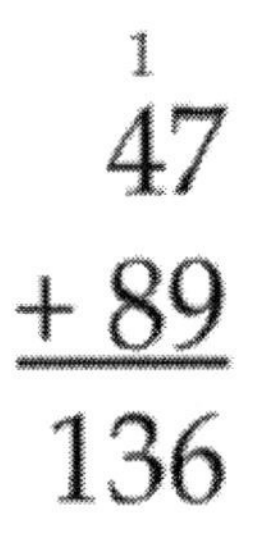

Let's look at adding 2 two-digit numbers. See the example at left. We always start with the *rightmost* column—the ones place. Think of it as adding up your $1 bills. We add 7 + 9 to get 16. Recall that we're only allowed to put one digit in any place.

Think about what 16 really means. If you were dealing with money, you could combine three $5 bills and one single, but remember that in math we only work with ones, tens, hundreds and so forth.

We'll have to treat the 16 as 6 ones and 1 ten. The 6 goes in the sum in the ones column, and the 1 ten gets **carried** into the tens column. We're reminding ourselves that when we add up the numbers in the tens column, we still have 1 ten which is "on hold" for the moment.

Now we move to the left to the tens place. We always make our way from right to left, to higher and higher place values. We add 4 + 8 to get 12, plus the 1 ten that we carried into that column, giving us 13 tens.

Think about what that really means. If you have thirteen $10 bills, you have $130. However, we still can't fit two digits into one column. The 130 is comprised of 3 tens, and 1 hundred. We can put the 3 tens in the tens column sum, but then we'll have to write the 1 hundred in its own column to the left. Had there been any digits to add in the hundreds column, we would have carried the 1 into the hundreds column just like we carried something into the tens column.

It's important that you not just do these steps mechanically, but that you keep track of what each column represents, and why the procedure works as it does.

TWO-DIGIT SUBTRACTION WITH BORROWING

Let's look at subtracting by hand to reinforce some basic concepts. See the example of 93 – 15. As with addition, we always start on the right in the ones place. Furthermore, we always must compute "top number minus bottom number." In this case, the situation in the ones place is 3 – 5.

This brings up an important point. We're not permitted to just simply reverse the numbers to make it 5 – 3 for an answer of 2. It just doesn't work that way. We must always do "top number minus bottom number." Don't read on until you accept that.

In a sense, we "cannot do" 3 – 5. In Chapter Four you'll learn that the answer to 3 – 5 is -2 (negative 2). The trouble is that in a subtraction problem such as the example, we must make sure that we do not end up with a negative answer in any column.

That brings us to the concept of **borrowing**. If you accept the fact that computing 3 – 5 will give us a negative answer, and if you also accept that we cannot take it upon ourselves to reverse the 3 and the 5, you'll see that we must somehow increase the 3 so that we'll have a larger number from which to subtract the 5.

We accomplish that by "borrowing." Remember that 1 ten is equal to 10 ones, just like a $10 bill is equal to ten singles. Look at the 9 in the tens column of the top number. There is no harm in "borrowing" one of those tens as long as we immediately give it back to the top number in the form of 10 ones. The 9 becomes 8, reducing the value of the top number by 10 for the moment. Then we put a little 1 next to the 3 in the ones place, effectively making it into 13, and giving back the 10 that we borrowed. We "cheated" in a sense by "squeezing" two digits into a place that we know is only allowed a single digit.

With that borrowing maneuver, we now have a larger number from which to subtract the 5. We compute 13 – 5 to get 8 which goes in the ones place of the answer. Moving left to the tens place, we compute 8 – 1 to get 7 in the tens place of our answer.

It's important to understand that we didn't actually break any math rules, and if you were to perform this subtraction problem using actual counters, you would get the same answer. You can also check your answer by adding 78 + 15 to see that you will get back to 93.

TWO-DIGIT BY ONE-DIGIT MULTIPLICATION

It's important to understand the math behind the basic procedure for multiplying numbers by hand. Many important concepts are reviewed in the process.

$$\begin{array}{r} {}^{5} \\ 28 \\ \times\ 7 \\ \hline 196 \end{array}$$

In this example of 28 × 7, we'll work in two stages. First, the 7 ones are going to be multiplied by the 8 ones. As with addition, we will carry the portion of the answer that belongs in the tens place. We'll then multiply the 7 times the 2, keeping in mind that the 2 is really 20 since it's in the tens place.

Let's start with multiplying our ones times our ones. We multiply 7 × 8 to get 56. As with addition, we must treat that as 5 tens and 6 ones. We'll carry the 5 into the tens place, not to multiply it, but to remember that later we must add it to our tens. You could think of the 5 tens as a $50 bill that has been set aside for the moment.

We then multiply the 7 ones times the 2 in the tens place of the first number. We're essentially **distributing** the 7 over both the 8 and the 2, which is discussed in the next chapter. 7 × 2 is 14, but we must remember that the 2 is really 20 as described above. We're computing 7 × 20 to get 140 which is 1 hundred and 4 tens.

We'll take those 4 tens and add them to the 5 tens that we carried earlier. That gives us 9 tens which goes into our answer in the tens place. We have 1 hundred left over which we can just write directly into the hundreds column of the answer. Had there been any numbers in that column, we would have had to carry that 1 and remember to add it in after multiplying whatever was in that column.

SO NOW WHAT?

If you're looking for information about long division or more complicated arithmetic with other operations, please see the Introduction for why those topics are omitted from this book.

Before progressing to the next chapter, it is essential that you fully understand all of the concepts in this one. Take time to review the material. See the Introduction and Chapter Twelve for information about how to get help, ask questions, or test your understanding.

CHAPTER THREE

Basic Math Topics and Operations

WHAT IS AN INTEGER?

The term **integer** is used very frequently in math, so you must memorize its definition. An integer is just a whole number. It can be positive, negative, or zero. Examples of integers are -17, 0, 32, and six million. Examples of numbers that are not integers are 3.7 and ½. You'll learn about non-integers later in the book.

EVEN AND ODD NUMBERS

If a number ends in 0, 2, 4, 6, or 8, we say that it's **even**. This is because it can be cut in half into two equal parts. Some examples of even numbers are 38, 164, 2, and 50. All that matters is the *rightmost* digit.

If a number ends in 1, 3, 5, 7, or 9, we say that it is an **odd** number. This is because it cannot be divided exactly in half. One part will have one more item in it than the other. Some examples of odd numbers are 49, 857, 1, and 23. All that matters is the *rightmost* digit.

Some problems are designed to see if you understand how even and odd numbers work when adding, subtracting, or multiplying different combinations of them. Questions involving division are usually avoided since division often results in an answer that is not a whole number, and therefore is neither even nor odd.

These questions are very easy if you don't make a big deal out of them. There are certainly sets of rules that can be memorized, but there is no point in doing so since it only takes a moment to determine the answer to these questions when they come up.

When a question refers to even and odd numbers, simply use 2 as a sample even number, and 1 as a sample odd number. For example, the question may ask if the result of multiplying two odd numbers is even or odd. Note that $1 \times 1 = 1$ which is odd, so that is the answer. It works that way for any two multiplied odd numbers.

As another example, if a question asks for the result of adding an even plus an odd number, just use $2 + 1$ as an example. The answer is 3, which is odd. It will work that way for the sum of any even and odd number.

There are far more important rules in math to memorize than rules about how even and odd numbers work. Just make up a quick, simple example whenever necessary.

GREATER THAN AND LESS THAN

We often need to determine whether one quantity is greater than or less than another. We use the symbols ">" and "<" for this purpose. The first symbol ">" is read as "**greater than**." It means that the number on the left is larger than the number on the right. For example, we can symbolically write $5 > 3$ since 5 is larger than 3. We read the statement from left to right.

The symbol "<" is read as "**less than**." It means that the number on the left is smaller than the number on the right. For example, we can symbolically write $7 < 19$ since 7 is smaller than 19. We read the statement from left to right, although we could actually read it in reverse to mean that 19 is greater than 7, which is also true.

It can be helpful to visualize the symbol as an alligator's mouth which is opening toward the larger quantity. Later we'll learn how to compare numbers that are non-integers such as fractions and decimals. The concept will be exactly the same, though.

INTRODUCING EXPONENTS (POWERS)

An **exponent**, sometimes referred to as a **power**, is a number written as a superscript to the right of another number which is known as the **base**. The exponent tells us how many copies of the base we will *multiply* times each other, including the original. It is just a shortcut notation. For example, 3^4 means to take the base of 3 and multiply it times itself using a total of four copies, including the original. To compute the answer we must multiply 3 × 3 × 3 × 3 which is equal to 81.

$$3^4 = 3 \times 3 \times 3 \times 3 = 81$$

"Three to the power of 4" or "Three to the fourth power"

There are several important points to be made. First, remember that exponents are a shortcut for *repeated multiplication*—not repeated addition. Also, understand that 3^4 does *not* mean 3 × 4. It means to take the base, and repeatedly multiply it times itself the specified number of times, including the original base itself.

Note that 3^4 can be read as "three to the power of 4," or "three to the fourth power," or just "three to the fourth." It is not read as "three four" since that is ambiguous, and it is certainly not read as "three times four," since that is not at all what it means. It also has nothing to do with the number 34 since the 4 is an exponent.

SQUARE, CUBE, AND OTHER SPECIAL POWERS

There are two exponents which are so common that they have special names. Those names have to do with topics in geometry which you'll learn in later math.

The first special exponent is 2. If a base is raised to the power of 2, we usually read the exponent as "squared." For example, 7^2 is usually read as "seven squared," but it would not be wrong to read it as "seven to the power of 2," or "seven to the second power."

To review, the notation means to take the base of 7 and multiply two occurrences of it together. We compute 7×7 to get 49. We can say that 49 is the **square** of 7. Symbolically we can say that any number to the power of 2 (or "squared") is equal to itself times itself.

$$n^2 = n \times n$$

Any number squared equals itself times itself

The next special exponent is 3. If a base is raised to the power of 3, we usually read the exponent as "cubed." For example, 8^3 is usually read as "eight cubed," but it would not be wrong to read it as "eight to the power of 3," or "eight to the third power." To review, the notation means to take the base of 8 and multiply three occurrences of it together. We compute $8 \times 8 \times 8$ to get 512. We can say that 512 is the **cube** of 8. Symbolically we can say that any number to the power of 3 (or "cubed") is equal to itself times itself times itself.

$$n^3 = n \times n \times n$$

A number cubed equals itself times itself times itself

Think about what happens if we raise a base to the power of 1. The exponent tells us how many copies of the base we must multiply together, including the original base. If we write 9^1, we're indicating to take the base of 9, but don't multiply it by any other occurrence of it—just leave it as is. It may seem moot, but 9^1 is simply 9. In later math you'll learn why such a rule is important. Symbolically we can say that any number to the power of 1 (or to the first power) is equal to itself.

$$n^1 = n$$

Any number to the power of 1 equals itself

What happens if you raise a base to the power of 0? It's tempting and reasonable (yet wrong!) to say that the answer is 0 with the logic that if you have no copies of a number, you have nothing no matter what you do.

Despite this logic, any* number raised to the power of 0 is actually equal to 1. The explanation is best left for the study of algebra. Until then, just remember that any* number raised to the power of 0 is equal to 1.

$$n^0 = 1$$

Any* number to the power of 0 equals 1

* 0^0 *is a special case that won't come up until later math.*

WHAT IS A PERFECT SQUARE?

A **perfect square** is a number which is the square of a whole number. For example, 49 is a perfect square because it is the square of 7 ($7^2 = 49$). The number 24 is not a perfect square because there is no whole number that we can square in order to get 24—the square of 4 will be too low, and the square of 5 will be too high. The reason why we use the term "perfect square" will make more sense when you later study geometry.

It's very important to memorize the perfect squares between 1 and at least 144 since they come up so often in math. When you see a number in that range, you should be able to instantly recognize if it is a perfect square, and if so, what number it is the square of.

The chart below shows the squares of numbers between 1 and 15, as well as some common others. It is important to memorize at least the first two columns of the chart, if not the entire chart. This will serve you throughout your study of basic math, and later algebra and geometry.

$1^2 = 1$	$5^2 = 25$	$9^2 = 81$	$13^2 = 169$	$25^2 = 625$
$2^2 = 4$	$6^2 = 36$	$10^2 = 100$	$14^2 = 196$	$30^2 = 900$
$3^2 = 9$	$7^2 = 49$	$11^2 = 121$	$15^2 = 225$	$40^2 = 1600$
$4^2 = 16$	$8^2 = 64$	$12^2 = 144$	$20^2 = 400$	$50^2 = 2500$

A table of common perfect squares to memorize

THE SQUARE ROOT OF A NUMBER

The **square root** operation is the inverse (opposite) of the squaring operation. It can be a bit tricky, so read this section slowly and carefully. If we are asked to find the square root of a number, the question is actually, "What number can we square in order to get the number in question?" We symbolize square root with the $\sqrt{\ }$ symbol. The number that goes under the bar is the number that we are being asked to find the square root of.

Let's look at an example and compute $\sqrt{64}$. This is read as "the square root of 64." The question is, "What number can we square in order to get 64?" The answer to that question is 8. The reason why is because if we square 8 (that is to say compute 8^2), we get 64.

It's important to understand that the answer to the above problem is 8—just plain 8. It is completely incorrect to answer 8^2. The question is asking for the number which would need to be squared in order to get 64. The answer is just 8. Certainly if you were asked to justify your answer, you could do so by noting that $8^2 = 64$, but that is not the answer. The only correct evaluation of $\sqrt{64}$ is 8. This is not a matter of "nitpicking" as many students believe. Any answer besides 8 will be marked wrong.

It's also completely incorrect to answer $\sqrt{8}$ since that is actually asking the question, "What number must we square in order to get 8?" Such a number does not actually exist, at least not a whole number, and it has nothing at all to do with the original question.

Below is a chart of square roots which evaluate to numbers between 1 and 50.

$\sqrt{1} = 1$	$\sqrt{36} = 6$	$\sqrt{121} = 11$	$\sqrt{400} = 20$
$\sqrt{4} = 2$	$\sqrt{49} = 7$	$\sqrt{144} = 12$	$\sqrt{625} = 25$
$\sqrt{9} = 3$	$\sqrt{64} = 8$	$\sqrt{169} = 13$	$\sqrt{900} = 30$
$\sqrt{16} = 4$	$\sqrt{81} = 9$	$\sqrt{196} = 14$	$\sqrt{1600} = 40$
$\sqrt{25} = 5$	$\sqrt{100} = 10$	$\sqrt{225} = 15$	$\sqrt{2500} = 50$

A table of common square roots to memorize

The chart is really the inverse of the chart of perfect squares. As with the perfect squares, it's important to memorize this chart. However, once you've memorized the chart of perfect squares, you essentially have this chart memorized as well since it is really just the inverse.

SQUARING AND "SQUARE ROOTING" ARE INVERSE OPERATIONS

As mentioned, it's important to understand that squaring and "square rooting" are inverse operations. They "undo" each other, just like subtraction undoes addition, and division undoes multiplication as we've seen. If you square a number, and then take the square root of that result, you will get back the original number. For example, if you compute 13^2 you get 169. If you then evaluate $\sqrt{169}$ you get 13 which is what you started with.

ORDER OF OPERATIONS (PEMDAS)

Many mathematical **expressions** (combinations of numbers and operations) combine several different operations. Some expressions include the same operation more than once, and some include portions within parentheses. At first glance such expressions can seem quite overwhelming such as the next example.

$$2000 - 100 \div 5^2 - 7 + 8 \times [\sqrt{81} + (30 \div 6)]$$

A sample "order of operations" problem

Rules have been established which dictate the order in which we must **evaluate** an expression, meaning how we go about simplifying it until we have one number. These problems are actually easy as long as you fully understand the rules (the **order of operations**), apply them correctly, and take one step at a time without combining several steps into one. It is also a matter of understanding the nuances of some of the rules, many of which are either not taught, or are insufficiently stressed.

The general idea is that the acronym **PEMDAS** (pronounced as written) reminds us of the order in which we must perform the various operations in an expression. As you'll see, though, the acronym is potentially somewhat misleading. First let's look at the order of operations, and then we'll look at some examples.

First Priority: Parentheses (left to right if more than one pair; inner to outer if nested)

The first step is to evaluate any portion of an expression in parentheses, if any. Ignore everything else until that is done. If there are parentheses nested within other sets of brackets or braces, we work our way from inner to outer. For example, even though the expression $2 \times (3 + 4)$ starts with multiplication, we first do what is in parentheses. That gives us 7, leaving us with 2×7, which is 14.

As another example, in the expression $5 + [\ 4 \times (2 + 1)\ ]$, first evaluate what's in the the innermost set of parentheses, then remove them. We now have $5 + [4 \times 3]$. Now evaluate what's in the remaining set of brackets, then remove them. We have $5 + 12$ which is 17.

In the expression $8 + (7 \times 3) + (12 \div 6) + 5$, first evaluate what is in the leftmost set of parentheses, then the rightmost, then proceed with adding as usual. The expression simplifies to $8 + 21 + 2 + 5$ which is 36.

Second Priority: Exponents (or square roots)

After evaluating and removing all of the parentheses (if there were any), the next thing to look for is exponents or square roots. If any are present, they must be evaluated before moving ahead to the next step. If more than one exponent and/or square root are present, just evaluate them in the order that they appear from left to right.

For example, even though the expression $3 + 5^2$ is really an addition problem, it is totally wrong to start by adding $3 + 5$. Seeing that there are no parentheses to deal with, we start by addressing the exponent, which in this case is 2, or squaring. We now have $3 + 25$ which is 28.

The expression $3 + \sqrt{100} + 2^4$ contains both a square root and an exponent. Evaluate them in the order they appear from left to right. After evaluating the square root we have $3 + \;\; 10 + 2^4$. After evaluating 2^4 we now have 3 + 10 + 8 which we add as usual to get 21.

Third Priority: Multiplication and Division (left to right, remembering that division could come first)

After handling parentheses (if any), and after handling any exponents or square roots (if any), the next step is to handle multiplication and division. This is done before any addition or subtraction is handled, even if addition or subtraction appear earlier in the expression than multiplication or division.

It's important to understand that we handle multiplication and division *from left to right* in the expression, keeping in mind that division could come first. It is wrong to resolve multiplication before division if the division appears to the left of the multiplication.

For example, in the expression $3 + 8 \times 9$, even though addition comes first, the multiplication has a higher priority. We must evaluate 8×9 to get 72. We now have $3 + 72$ which we can add to get 75. If we incorrectly did the addition first we would then have to multiply 11 × 9 for an answer of 99 which would be wrong.

As another example, in the expression 12 ÷ 3 × 2, we have both multiplication and division. Whenever an expression is reduced to (or starts out with) nothing but multiplication and division, we do those operations in order from left to right, in the order that they appear. We do not necessarily do the multiplication before doing the division, despite many students' notions to the contrary.

In the above example, we first compute 12 ÷ 3 to get 4, leaving us with 4 × 2 which is 8. If we incorrectly did the multiplication first, we would be left with 12 ÷ 6 for an answer of 2 which would be wrong.

Fourth Priority: Addition and Subtraction (left to right, remembering that subtraction could come first)

Although multiplication and division share the same level of priority, and are evaluated left to right, it is still the case that multiplication and division are always performed before addition and subtraction. We don't do addition and subtraction until the expression has been simplified to the point where it has nothing else.

Once an expression has nothing remaining but addition and subtraction, we do the addition and subtraction in order from left to right, in the order that they appear. We don't necessarily do the addition before the subtraction, despite many students' notions to the contrary.

In the expression 10 – 2 + 3, we simply work from left to right since we have nothing but addition and subtraction. In this case the subtraction happens to come first. We compute 10 – 2 to get 8, and now we have 8 + 3 which is 11. If we incorrectly did the addition first, we would be left with 10 – 5 for an incorrect answer of 5. Remember that addition and subtraction have the lowest priority when evaluating an expression. Don't be tempted to do those operations first in order to "get them out of the way." That will lead to the wrong answer.

Summary: So what is PEMDAS?

PEMDAS is an acronym that helps us remember the general concept of the order of operations. From left to right the letters stand for **P**arentheses, **E**xponents, **M**ultiplication, **D**ivision, **A**ddition, and **S**ubtraction.

As we've seen, this is somewhat misleading. Multiplication and division share the same priority. While they are always done before addition and subtraction, it's not necessarily the case that multiplication is done before division, just because M comes before D in PEMDAS. They are done from left to right, in the order that they appear. Sometimes division will come first.

The same is true with addition and subtraction. Although they are the last operations to be done when simplifying an expression, they share the same priority. It's not necessarily the case that addition is done before subtraction, just because A comes before S in PEMDAS. They are done from left to right, in the order they appear. Sometimes subtraction will come first.

As you work through PEMDAS problems, just handle them one step at a time. Write the results of each step on its own new line, and draw arrows to remind yourself of how you got from one line to the next.

THE DISTRIBUTIVE PROPERTY OF MULTIPLICATION OVER ADDITION AND SUBTRACTION

Let's look at the expression $2 \times (3 + 5)$. In the last section we learned that we must first first evaluate whatever is in parentheses. Normally multiplication would have a higher priority than addition, but in this case the addition is in parentheses which has the highest priority. We evaluate the addition to get 8, leaving us with 2×8, or 16.

There is an important rule in math which states that in an expression like this, we can "**distribute**" the multiplication over the addition which is in parentheses. We can do this by having the 2 multiply the 3, and then having it multiply the 5, and then adding both of those products together. The problem becomes (2 × 3) + (2 × 5). The first set of parentheses evaluates to 6, and the second set evaluates to 10. Notice how we get 16 when we add them, which is the same answer we got above.

At this point many students ask what the point of distributing is. One answer is that we can use the property when we want to multiply a number that is somewhat "complicated" on its own, but that is "simpler" after we split it up in some way. Think about multiplying 23 × 101. That would be difficult to do in your head using the traditional multiplication procedure, and even if you did it by hand it would require several lines worth of work.

Let's instead change the problem so that we first "break up" 101 into (100 + 1). There is no harm in that since 100 + 1 certainly does equal 101. Now, using the distributive property of multiplication over addition, we can first easily multiply 23 times 100 to get 2300 (just tack on 2 zeroes), and then we can multiply 23 times 1 to get 23. Add the products of 2300 and 23 together to get 2323.

$$\begin{aligned} 23 \times 101 &= 23 \times (100 + 1) \\ &= (23 \times 100) + (23 \times 1) \\ &= 2300 + 23 = 2323 \end{aligned}$$

Using the distributive property to simply a computation

While the distributive property may not seem very important at this point, it will be extremely important later when you study algebra, so it is best to become comfortable with it at this point. Note that the property can be applied to subtraction in the same way that it applies to addition. It is symbolized below for any numbers *a*, *b*, and *c*.

$$a \times (b + c) = (a \times b) + (a \times c)$$
$$a \times (b - c) = (a \times b) - (a \times c)$$

The distributive property of multiplication over addition and subtraction

WHAT IS A FACTOR?

The term **factor** is best defined by example. The factors of 12 are 1, 2, 3, 4, 6, and 12. All of those numbers divide into 12 evenly, that is to say with no remainder. Recall that division and multiplication are inverse operations. In this example, that means that since 6 is a factor of 12, there must be some number that we can multiply times 6 in order to get back to 12. That number of course is 2, which means 2 is also a factor of 12.

Factors always come in pairs like we just saw. In the example above, 1 is paired with 12, 2 is paired with 6, and 3 is paired with 4. Sometimes students draw arcs to connect the members of each pair, just to be certain that no factors were omitted from the list.

Numbers that are perfect squares also have their factors occur in pairs, but the middlemost factor in the list will simply be paired with itself. For example, 6 is a factor of 36, but 6×6 makes 36, so there is no other number for 6 to be paired with.

It's important to understand that every number except 1 always has at least two factors—1 and itself. Every number can be evenly divided by 1 and itself. The reason why 1 is an exception is because even though 1 can be divided by 1 and itself, those are both the same numbers. Therefore 1 has just one factor, namely 1.

PRIME AND COMPOSITE NUMBERS

A number is called **prime** if it has two *unique* factors—1 and itself. Recall that the factors of a number are the numbers which divide into it evenly with no remainder.

For example, the factors of 24 are 1, 2, 3, 4, 6, 8, 12, and 24. That means that 24 is certainly not a prime number since it has many factors other than 1 and itself (24).

The study of prime numbers is important at all levels of math. However, unless you choose to study advanced math, the only thing you will likely be asked is whether or not a particular number is prime.

Think about the number 17. Its only factors are 1 and itself (17). There are no other numbers that divide into it evenly. That means that 17 is a prime number.

Think about the number 2. It only has two factors: 1 and itself (2). Nothing else divides into it evenly, at least no other whole number which is all that we're concerned about. That means that 2 is a prime number.

Are there any even prime numbers besides 2? Other even numbers will have 1 and themselves as factors, but will also have 2 as a factor since they're even numbers. That means that no even number besides 2 is prime. All other even numbers will have, at the very minimum, 1, 2, and themselves as factors, which means they aren't prime.

Numbers that are not prime are called **composite**. Some examples are 36, 100, and five million. Each has many factors besides 1 and itself. For example, the number 27 has factors of 3 and 9 in addition to 1 and itself, so it's composite. The number 4 is composite because it has an additional factor besides 1 and itself, namely 2.

What about 1? Its only factor is itself. It seems like it fits the definition of prime since its factors are 1 and itself, but those two factors are not unique. That makes 1 a unique case. It is neither prime nor composite.

THE PLACE VALUE CHART UP TO BILLIONS

We've already worked with numbers into the thousands. As we've seen, the place value chart is based on groupings of 10. Each column is worth 10 times the value of the column on its right. Reading from right to left we have ones (units), tens, hundreds, and thousands.

We can continue this pattern to extend the chart to accommodate larger numbers. To the left of the thousands place we have **ten thousands** which is ten times the thousands place. To the left of that we have **hundred thousands** which is ten times as large as ten thousands. To the left of that we have, in a sense, "thousand thousands" which is ten times as large as hundred thousands.

We have a special name for that place known as **millions.** One million is a thousand thousand (1000 × 1000).

Continuing the pattern we have places for **ten millions, hundred millions**, and **billions**. A billion is one thousand million. Next we have ten billions, hundred billions, and **trillions**. A trillion is one thousand billion—enough to discuss America's national debt until it reaches into the **quadrillions.** There is no such thing as a "zillion". It's just a fictitious, large number.

Billions	Hund. Millions	Ten Millions	Millions	Hund. Thousands	Ten Thousands	Thousands	Hundreds	Tens	Ones
7	2	3	4	0	5	6	1	8	9
7,000,000,000	200,000,000	30,000,000	4,000,000	0	50,000	6,000	100	80	9

Place value breakdown of the number 7,234,056,189
(Seven billion, two hundred thirty-four million, fifty-six thousand, one hundred eighty-nine)

When we write large numbers, we place a comma in between every three digits as we count starting from the right. This is just to make large numbers easier to read. The bold lines in the chart below indicate where the commas go, although we sometimes omit the comma for numbers that don't go beyond the thousands place.

READING LARGE NUMBERS WRITTEN IN WORDS

It's important to be able to read large numbers written in words. Look at previous caption as an example. We treat each grouping of three columns as its own entity.

As another example, we write 234 as "two hundred thirty-four," almost as if the entire number was in the units place. We write 567,000 as "five hundred sixty-seven thousand," almost as if the "567" part was entirely in the thousands place. The pattern continues with the digits that span the three columns of the millions grouping, and so on. The bold lines in the place value chart separate the main groupings of ones, thousands, millions, and billions.

Note that we don't use the word "and" when we use words to write whole numbers (numbers with no decimal component). In Chapter Nine we'll learn that we use the word "and" to represent the decimal point.

ROUNDING NUMBERS TO VARIOUS PLACES

Sometimes it's easier to work with an approximate instead of an exact value of a number. This is especially true when you are trying to estimate the answer to a problem which is an excellent habit to develop. For example, if you are asked to multiply 796 × 51, it's a good idea to first **round** each number in order to convert it to the simpler problem of 800 × 50, which you can compute as an estimate.

What you're doing is converting each number to a "simpler" number that is "close" to the original. The answer to the modified problem is an estimate of the answer to the original one. If your answer to the original problem is not close to your estimate, you'll know that you did something wrong.

In problems which ask you to round a number in some specific way, you will be told to what place you must round the number, for example, to the nearest hundred or nearest thousand.

There is a simple procedure to follow. For example, let's round 24,739 to the nearest hundred. Let's call the hundreds place our "target place," and bold it (see next page). To round to a given place, we always examine the place to the right of the target place. Let's call the digit in the place on the right the "check digit," and underline it. In the example we have 7 as the target digit, and 3 as the check digit.

Here is the first rule of rounding. If the check digit is 4 or lower, the digit in the target place *remains as is*, the check digit becomes a 0, and all the digits to the right of it become 0. In this example, 24,739 rounded to the nearest hundred is is 24,700—we rounded down.

$$24{,}\mathbf{7}\underline{3}5 \rightarrow 24{,}700$$

Rounding 24,735 to the nearest hundred

Now let's round that same number to the nearest thousand. We now have 4 in the target place, and 7 as the check digit.

Here is the second rule of rounding. If the check digit is 5 or higher, the digit in the target place is increased by 1, and the check digit and all the digits on its right become 0. In this example 24,739 rounded to the nearest thousand is 25,000—we rounded up.

$$2\mathbf{4}{,}\underline{7}35 \rightarrow 25{,}000$$

Rounding 24,735 to the nearest thousand

Let's round 39,512 to the nearest thousand. The target place digit is 9, and the check digit is 5. Since the check digit is 5 or higher, we must increase the digit in the target place by 1, and make the check digit and everything to the right of it 0. But in this case, how do we increase the target place digit of 9? Since we can't make it 10, we make it 0 and effectively carry the 1 so that the 3 is increased to 4. In this example, 39,512 rounded to the nearest thousand is 40,000—we rounded up.

$$3\mathbf{9},\underline{5}12 \rightarrow 40{,}000$$

Rounding 39,512 to the nearest thousand

SO NOW WHAT?

Before progressing to the next chapter, it is essential that you fully understand all of the concepts in this one. If you do not, you will have tremendous difficulty with later work, and will need to return to this chapter. Take time now to review the material. See the Introduction and Chapter Twelve for information about how to get help, ask questions, or test your understanding.

CHAPTER FOUR

Working with Negative Numbers

WHAT IS A NEGATIVE NUMBER?

There are many different ways to think about negative numbers. One way is to think of positive numbers as quantities that you *have* (assets), and negative numbers as quantities that you *owe* (debts). This is typically how information is displayed on a checking account statement. Deposits are represented as positive dollar amounts—money that you have. Checks you've written are represented as negative amounts—money that you owed. The positive amounts are offset by the negatives.

If you have more money than you owe, you will have a positive balance. This is a "good" situation, represented by a positive number. A positive balance tells you how much money you still have after paying all of your debts. If you owe more than you have, and have written checks for more money than is in your account, you will have a negative balance. This is a "bad" situation, represented by a negative number. It shows how much money you actually owe to the bank.

THE NEGATIVE SIDE OF THE NUMBER LINE

Negative numbers are represented to the left of 0 on the number line. They form the mirror image of the positive numbers on the right, except that they are each preceded by a negative sign (–). As we start to do basic arithmetic involving negative numbers, it is important to always visualize both sides of the number line.

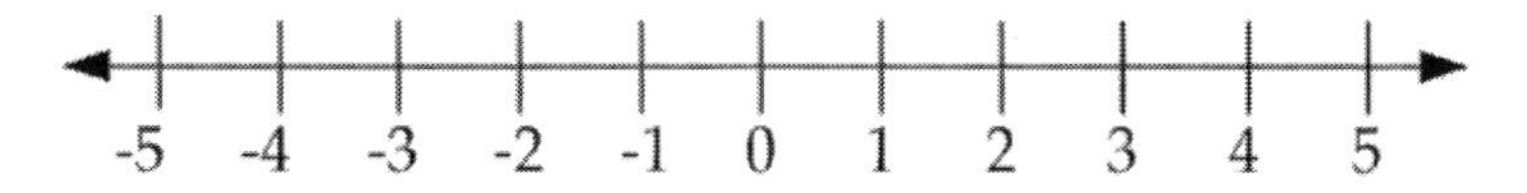

A number line that includes negative numbers

We refer to positive and negative numbers collectively as **signed numbers**. Negative numbers are always preceded by a negative sign—typically a short dash like this: (–). Usually we don't put a plus sign (+) in front of positive numbers since the absence of a negative sign implies they are positive. However, a plus sign is sometimes included for clarity. We also sometimes optionally write signed numbers in parentheses such as (-4) and (+5).

ABSOLUTE VALUE

The **absolute value** of a number tells us its distance from 0 on a number line. It may be on the left or right of 0, but measures of distance are always positive.

The absolute value of a number is simply the positive version of that number. For example, the absolute value of 3 is simply 3. It started out positive, and it stayed that way. The absolute value of -4 is 4. The answer is the positive version of the given number.

Absolute value is such an easy concept that students are often insulted by its simplicity. Still, it plays an important role in signed number arithmetic, and a more advanced role in later math.

ADDING SIGNED NUMBERS

There are many different ways to think about adding signed numbers such as (-4) + 7, but the math behind each is the same. The following steps offer a simple and effective way of thinking about how to add signed numbers, but may be somewhat different than what you've been taught or what is described in another book.

We have already learned how to add positive numbers to positive numbers. You can think of it as adding up your assets—quantities you have. We now need to learn how to add positive numbers to negative numbers, and how to add negative numbers to negative numbers.

ADDING A POSITIVE PLUS A NEGATIVE NUMBER

Recall that addition is commutative, meaning the order in which we add two numbers doesn't matter. Let's think about the case of adding a positive and negative number, for example, (-4) + (7). Think about what these numbers can represent. A positive number can be thought of as a quantity that you have, so imagine that you have $7. A negative number can be thought of as an amount that you owe. In this case, imagine that you owe someone $4.

Imagine we're balancing our checkbook by adding up the entries. The 7 represents money we deposited, and the -4 represents a check we wrote to pay off our debt.

When doing signed number addition of numbers with opposite signs, the first question to ask yourself is, "Do I have more money than I owe, or vice-versa?" In this case we have more money than we owe, which is a "good" or "positive" thing. Our answer will definitely be positive.

The second question to ask is, "How much more do I have than I owe?," or vice-versa if you owe more than you have. In this case we have $3 more than we owe. We represent our answer as positive 3. We write (-4) + 7 = 3.

What we really did to get the answer was subtract to find the difference between the absolute values (positive versions) of each number. We ignored the signs of each number for a moment, and computed the difference between them. There was a $3 difference between the amount we had, and the amount we owed. Since we had more than we owed, our answer was positive.

Let's try the reverse situation with (5) + (-9). We're adding two signed numbers with opposite signs. The first thing we must determine is if we have more than we owe. We do not. We owe more than what we have. That is a "bad" situation, so our answer will be negative.

Now let's find the difference between what we have and what we owe. Remember, we'll only examine the absolute values of the numbers, temporarily ignoring their signs. In doing this we can subtract to see that we owe $4 more than what we have. The best we can do is pay back the $5 that we have, and we'll still owe $4. We will represent that "bad" scenario with a negative answer, showing that we're still in debt. We write (5) + (-9) = -4.

SUMMARY OF ADDING SIGNED NUMBERS WITH OPPOSITE SIGNS

First determine if you have more than you owe. If you do, the answer is positive. If you owe more than you have, the answer is negative. Next, ignore the signs of each number for the moment, and subtract to determine how much more you have than you owe, or vice-versa. Those two pieces of information comprise your answer. For example, (-8) + (6) = -2, and (14) + (-9) = 5.

ADDING TWO NEGATIVE NUMBERS

Adding two negatives is easy. Think of it as adding up your debts. Your final answer will always be negative.

For example, let's say you owe one friend $30 and you owe another friend $50. Each of those numbers must be written with a negative sign to show that they are debts. To see how much you owe in total, you'll need to add those two numbers. Our problem is (-30) + (-50). Simply ignore the signs for a moment, and add the numbers themselves. We get 80 for our answer which we must write with a negative sign to show that it is a debt—it is what we owe. We write (-30) + (-50) = (-80).

Adding two negative numbers is always as simple as that. The answer will always be negative. Make up your own practice exercises involving adding two negative numbers. Just add the absolute values of the numbers, and then make your final answer negative.

REVIEW OF ADDING SIGNED NUMBERS

A **positive** plus a **positive** is always **positive**. Just add.

A **negative** plus a **negative** is always **negative**. Just add the absolute values of each number, ignoring the negative signs, and then make your answer negative.

To add a **positive** plus a **negative** number (recall that the order doesn't matter), think of the positive number as what you have, and the negative number as what you owe. Then perform the following two steps:

Step 1 of 2: If you have more than you owe, the sign of the answer is positive, and if not, it's negative.

Step 2 of 2: Ignore the signs of the numbers for the moment, then subtract to compute the difference between what you have and what you owe. Your final answer will combine the sign that you determined in Step 1, and the difference that you determined in Step 2.

Practice and memorize these procedures. Create and solve your own examples until you are convinced that all of them can be solved in the manner described above.

SUBTRACTING SIGNED NUMBERS (AN OVERVIEW)

We already covered the general idea behind subtraction in the chapter on basic arithmetic. In this section we'll look at how to handle subtraction problems that involve working with negative numbers. It can be confusing, but simple steps will be presented for solving such problems.

Let's take a look at the problem of 2 – 5 that was referenced earlier. As discussed, many students say that such a problem "cannot be done," or that it must be converted into 5 – 2. That is completely wrong.

Let's do the problem on a number line in the same way that we computed 7 – 3 earlier. To compute 2 – 5 on a number line, start at the first number, 2, and then count 5 tick marks to the left. Remember to count the 0 tick mark. We end up at -3, which is our answer.

Think about why that answer makes sense. Imagine we have $2, but we want to pay off a debt of $5 that we owe to someone. We're trying to take away more than we actually have. We cross into negative territory. We can pay the person back \$2, but we will still owe them \$3, represented by a negative number. Keep thinking about why it makes sense that 2 – 5 = -3, and that it is incorrect to change the exercise into 5 – 2.

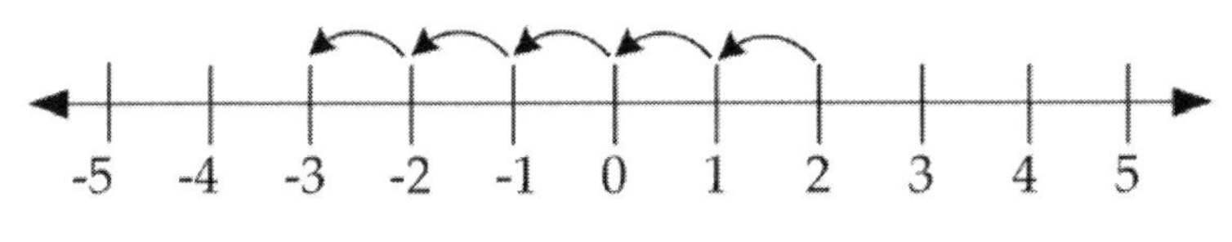

Computing 2 – 5 = -3 on a number line

Subtracting signed numbers can sometimes be difficult to translate into a real world situation. Also, as with addition, there are many different procedures you can follow, all of which are equivalent. The four-step procedure described on the next page may not match what you have learned or what is described in your textbook, but it is equivalent, and probably much simpler, and is guaranteed to always work.

If you follow the steps you will have absolutely no trouble, and you will get the correct answer. Difficulty will only arise if you either forget to follow the steps, or insist on solving the problem in some other way which may end up getting you confused.

HOW TO SUBTRACT TWO SIGNED NUMBERS

Step 1 of 4: Leave the first number alone. Don't touch it.

Step 2 of 4: Change the subtraction (minus) sign to an addition (plus) sign. This does not involve changing the sign of either number. It involves changing the actual operation of the problem from subtraction to addition.

Step 3 of 4: Change the sign of the second number to its opposite. If it was negative, make it positive. If it was positive, make it negative.

Step 4 of 4: You have converted the subtraction problem into an equivalent addition problem. You've already learned how to solve signed number addition problems in a previous section. Just follow the rules of that section to solve what is now an addition problem.

Example: To compute (-9) – (-7), leave the first number alone, change the subtraction operation to addition, and change the sign of the second number. We now have the addition problem (-9) + (7). Follow the rules in the section on signed number addition to get (-2).

That is all there is to it. Convert signed number subtraction problems into equivalent addition problems, then follow the rules for signed number addition.

Make up your own signed number subtraction problems in various combinations, and practice solving them using the four steps until you can do so easily. Resist the temptation to disregard or modify the steps.

DOES A DOUBLE NEGATIVE BECOME A POSITIVE?

Students are often taught the "shortcut" that in problems such as (4) – (-3), the "double negative" can be converted to a positive. This parallels the grammar rule for double negatives as in the sentence, "I don't have no money."

It is true that the two consecutive dashes in the problem above can each be converted to plus signs, giving us 4 + (+3), which of course equals 7. However, this rule is often applied incorrectly by students, and in situations where it simply is not valid. It is therefore avoided in this book. If you apply our four-step procedure to the above problem, it will still be converted to 4 + (+3). The advantage is that you will be using a procedure which always works, and is not based on any "shortcuts" that can be misunderstood or applied incorrectly.

IS IT A NEGATIVE SIGN OR A SUBTRACTION SIGN?

When doing signed number arithmetic, students often ask whether a particular dash is a negative sign or a subtraction sign. Of course the answer depends upon the particular context. However, it's important to understand that we always need an operation in an expression.

If we see something like (3 - 5), the dash must be taken to be a subtraction sign, even though it is short and printed non-centered between the two numbers. There is never a case when we'll have two numbers next to each other without them being connected by an operation (e.g., addition, subtraction, etc.) unless they are part of a list in which case they would be separated by a comma.

If we see something like (7 + – 4), we have to take the plus sign to be an addition operation (since it's not a comma-separated list, and there must be some operation), and we must take the dash to be a negative sign for the 4. No other interpretation makes mathematical sense.

Usually in print a subtraction sign is written with a longer dash than a negative sign, but this isn't always the case. It shouldn't be depended upon, nor should the spacing between numbers and symbols. Just refer to the previous examples to determine what makes sense.

ISSUES WITH SUBTRACTING ON A NUMBER LINE

Many students are taught how to do signed number subtraction on a number line like we saw in Chapter

Two. While there is nothing wrong with using a number line for "simple" subtraction problems such as 7 – 3, the situation can become convoluted when you try to use a number line to compute problems involving signed number subtraction such as (-7) – (-8). Certainly such a problem can be solved using a number line, but doing so involves being extra careful to not get confused.

It is far better to take the few moments necessary to follow the four steps outlined in this chapter to convert the subtraction problem into an addition problem, and then follow the simple rules for signed number addition. If you do this, you will always get the correct answer whereas when students attempt to solve such problems with other methods, it often becomes a matter of taking repeated incorrect guesses at the answer.

WHY "DIFFERENCE" IS A MISLEADING TERM

In Chapter Two we defined **difference** as the answer to a subtraction problem, and said that the term is somewhat misleading. In a sense, the difference between 3 and 7 is the same as the difference between 7 and 3. In everyday life, most people would that have a difference of 4, namely the distance between the two numbers.

The trouble is that from a math standpoint, such a statement implies that 3 – 7 is the same as 7 – 3 which is completely false, and the source of a great deal of confusion for many students.

Just remember that when we do a subtraction problem, we call the answer the difference, just as when we add, multiply, and divide, we call the answer the sum, product, and quotient, respectively. When we speak in terms of the "distance" or "span" between two numbers, we'll use the term **range** as discussed in Chapter Eleven.

MULTIPLYING SIGNED NUMBERS

Multiplying signed numbers is much easier than adding or subtracting them. There are just four simple rules. In each case, multiply the numbers involved, then apply the appropriate rule.

POSITIVE × **POSITIVE** = **POSITIVE**

POSITIVE × Negative = Negative

Negative × **POSITIVE** = Negative

Negative × Negative = **POSITIVE**

The rules for multiplying signed numbers

It is helpful to remember that when multiplying numbers with matching signs, the answer is positive ("matching is good"). When multiplying numbers with mismatched signs, the answer is negative ("mismatched is bad").

When we added signed numbers with opposite signs, it made a difference which number had the negative sign. We had to determine if we had more than we owed. That isn't an issue with signed number multiplication. Multiplying opposite signs always yields a negative answer.

You can think of multiplying opposite signs as repeated debt. For example, think of (-6) × (3) as a debt of $6 repeated three times. You owe $6 to each of three different people. You owe $18 in total, so the answer is (-18).

It is hard to find a "real-world" analogy for multiplying a negative times a negative. Just remember the rule that multiplying matching signs results in a positive answer.

DIVIDING SIGNED NUMBERS

The rules for dividing signed numbers are easiest of all. They are exactly the same as the rules for signed number multiplication. Just divide the numbers involved (ignoring the signs for the moment), then apply the same rules that are used for multiplying signed numbers.

POSITIVE ÷ POSITIVE = POSITIVE

POSITIVE ÷ Negative = Negative

Negative ÷ **POSITIVE** = Negative

Negative ÷ Negative = **POSITIVE**

The rules for dividing signed numbers

THE SQUARE OF A NEGATIVE NUMBER

A common point of confusion among students is the difference between the square of a negative number, and the negative of a number squared. They are not interchangeable. See the next example. We have -7 in parentheses with an exponent of 2 outside. Remember that according to PEMDAS, we always must first evaluate what is in parentheses. In this case, there is nothing to do within parentheses. We then move onto the exponent which affects the contents of the parentheses. We have (-7) × (-7) or 49.

$$(-7)^2 = (-7) \times (-7) = 49$$

Squaring a negative number

In the next example we don't have parentheses. Should we do our squaring first, and then make the answer negative, or do we square a negative number?

$$-7^2 = -(7^2) = -49$$

Making a squared number negative

The answer to that question involves knowing what it means to make a number negative. When we put a negative sign in front of a number, what we're really indicating is to multiply the number times -1. We'll work with that concept more later.

For the purposes of our problem, we have an exponent, and we have implied multiplication. According to PEMDAS, the exponent has higher priority. We first square the 7 to get 49, and then we make it negative with implied multiplication by -1, giving us an answer of -49.

THE SQUARE ROOT OF A NEGATIVE NUMBER

Until much later math, we say that the square root of a negative number is "**undefined**"—it "cannot be done." Remember that square root asks, "What number can we square in order to get this number?" Think about $\sqrt{-16}$. What number can we square in order to get -16?

Recall that squaring means to multiply a number times itself. When we do that we will always get a positive answer since the signs of the numbers match. We can't end up with a negative number.

$$\sqrt{-16} = Undefined$$

The square root of a negative number is undefined

POSITIVE NUMBERS HAVE TWO SQUARE ROOTS

Now that we're learning about negative numbers, let's take the concept of square root a step further. Remember again that square root asks the question, "What number can we square in order to get this number?" Think about $\sqrt{16}$. What number can we square in order to get 16?

We've already seen that 4 is certainly an answer. If we square 4, we get 16. That means that $\sqrt{16} = 4$. Are there any other solutions? It is wrong to say 8×2. Certainly those two numbers do multiply to 16, but they are not the same number. We need a number that we can multiply times itself in order to get 16.

Let's extend our search to the world of negative numbers. Remember that if we multiply a negative number times itself, we get a positive answer. That means that if 4 was an answer, it stands to reason that -4 would also be an answer. $(-4)^2 = 16$, so that means that $\sqrt{16} = -4$.

We say that the square root of 16 equals "plus or minus 4." There is a positive and a negative answer, and both are valid. This is written symbolically as $\mathbf{\sqrt{16} = \pm 4}$.

Even though there are two answers, we refer to the positive answer as the **principal** square root. This is the "default" answer we give. One reason is that square roots play a big role in geometry which you'll learn about later. Geometry deals with distance which is always positive. In such practical cases we "discard" the negative evaluation of a square root, even though it is mathematically valid.

$$\sqrt{16} = \pm 4$$

The square root of a positive number has "plus and minus" versions. The positive one is considered the "principal" root.

MORE ABOUT ABSOLUTE VALUE

We are sometimes asked to compute the absolute value of an expression, such as 4 – 6. That expression evaluates to -2, so its absolute value is 2. The absolute value of a number or an expression is always positive.

We use vertical bars around a number to represent the absolute value operation. For example, $|-7| = 7$. If there is an expression within the vertical bars, we must evaluate that expression first, before we apply the absolute value operation to it. Review the previous sections on signed number arithmetic, then study these examples:

$$|3| = 3 \qquad |-7 + 6| = 1 \qquad |(-3) \times (-5)| = 15$$

$$|-4| = 4 \qquad |10 - 8| = 2 \qquad |20 \div (-4)| = 5$$

Computing the absolute value of various expressions

SO NOW WHAT?

Before progressing to the next chapter, it is absolutely essential that you fully understand all of the concepts in this one. In particular, if you do not fully master how to perform the four basic arithmetic operations with signed numbers, you will run into endless difficulty with all of the math that you will study from this point forward. None of this is "busy" or "baby" work. It is the foundation of math.

Take as much time as you need to review the material in this chapter, and return to it as often as necessary until all of it becomes second nature to you, and you are no longer confused or intimidated by the sight of a negative number. See the Introduction and Chapter Twelve for information about how to get help, ask questions, or test your understanding.

CHAPTER FIVE

Basic Operations with Fractions (+, –, ×, ÷)

WHAT IS A FRACTION?

A fraction is simply a value that represents part of a whole. There are two key points that must be understood. First, when we speak about a fraction, we are referring to some whole thing in a general sense. Get out of the mindset of, "What if my pizza pie was bigger than yours in the first place?" While it is true that ½ of 500 apples is a larger quantity than ½ of 32 apples, that is not the point. For the purposes of our work with fractions, a fraction just refers to a part of some generic whole.

Whenever we divide a whole into parts, all of the parts must be equal in size. Get out of the mindset of, "What if the pie wasn't cut perfectly, and my slice was bigger than yours?" No matter what we're dealing with, we're always going to speak of some general whole, divided into a specified number of equally-sized parts.

All fractions have a top part and a bottom part, separated by a horizontal line. The number in the bottom part is called the **denominator**. It tells us how many *equally-sized* parts the whole has been divided into. For example, a pizza pie is typically cut into 8 equal slices. Fractions involving pizza typically have a denominator of 8. We can speak of how many *eighths* we ate. The reason why we say "eighths" and not "eights" is explained later.

$$\frac{\textit{numerator}}{\textit{denominator}}$$

The top number is called the **numerator**. It tells us how many of the parts we are concerned about. It could be how many parts we have eaten, or how many are a particular color, or whatever the problem deals with. For example, if you ate 3 slices of pizza, you could say that you ate 3/8 (three-eighths) of the entire pie. Note that throughout this book, fractions will sometimes be written in the form *a/b* instead of $\frac{a}{b}$.

THE EFFECT OF INCREASING / DECREASING THE NUMERATOR / DENOMINATOR OF A FRACTION

It is important to understand the effect of increasing or decreasing the numerator or denominator of a fraction.

For example, what happens to a fraction if we increase the denominator, but leave the numerator alone? Think about eating 3 slices of a pizza pie that has been cut into 8 equal slices. You ate 3/8 of the pie. Now imagine that the same original pie was cut into 20 equal slices instead of 8, and you ate 3 of those pieces. You ate 3/20 of that differently-cut pie.

Remember that in both cases we're starting out with the same size pie—we just cut the second pie into more (and therefore smaller) parts. It's easy to see that the second fraction of 3/20 represents less food. You still ate 3 parts, but the parts of the second pie were smaller. This shows that as the denominator of a fraction gets bigger, the actual value of the fraction gets smaller.

This scenario can be looked at in reverse. If the pie had been cut into only 4 slices and you ate three of them, you would have eaten 3/4 of the pie—a larger value than the original 3/8. This shows that as the denominator of a fraction gets smaller, the value of the fraction gets larger.

Manipulating the numerator of a fraction is easier to think about. The numerator of a fraction tells us how many pieces of a whole we are concerned about. Obviously, more parts of the same whole imply a larger quantity. Therefore, as the numerator of a fraction increases, the value of the fraction increases, and as the numerator decreases, the value of the fraction decreases.

ADDING FRACTIONS WITH LIKE (MATCHING) DENOMINATORS

Imagine that you ate one slice of a typically-cut pizza pie, and I ate four slices. Together we ate five slices. We can say that combined we ate 5/8 (five-eighths) of the pie. We don't add the denominators to get 16 since the pie didn't suddenly get cut into 16 slices. It's still a whole pie of 8 equal slices, of which you ate 1, and I ate 4.

That is how adding fractions works when we have like (matching) denominators. We just add the parts as represented by the numerators. We maintain the same denominator throughout the entire problem including the answer since the total number of parts hasn't changed at all. All we did was add the numerators.

$$\frac{1}{8} + \frac{4}{8} = \frac{5}{8}$$

Adding the numerators of fractions with like denominators

SUBTRACTING FRACTIONS WITH LIKE (MATCHING) DENOMINATORS

Subtracting fractions with matching denominators follows the same procedure, but we'll subtract instead of add. Imagine that I ordered five slices of pizza, but I only ate two of them. What fraction of the whole pie do I have left over. I started with 5/8 of the whole pie, and 2/8 were taken away because I ate that part. Subtract the numerators to see that I have 3/8 of the original pie remaining.

Again, notice how the denominator remains constant throughout the problem. We don't subtract the denominators since we still have a pie of 8 equal slices.

$$\frac{5}{8} - \frac{2}{8} = \frac{3}{8}$$

Subtracting fractions with common denominators

ADDING AND SUBTRACTING FRACTIONS WITH UNLIKE DENOMINATORS, AND OTHER CONCERNS

Sometimes we end up with an answer in which the numerator is larger than the denominator, or an answer which can be "reduced" (simplified). Those situations are discussed in the next chapter. For now, just be sure to understand the procedure for adding and subtracting fractions with like (matching) denominators. In the next chapter we'll also learn how to add and subtract fractions that have unlike (non-matching) denominators.

MULTIPLYING A FRACTION TIMES A FRACTION

Multiplying fractions is easy as long as you memorize one simple rule, and don't get it confused with anything else. To multiply two fractions, just multiply *straight across* the numerators to get the numerator of the product, and *straight across* the denominators to get the denominator of the product. It doesn't matter whether or not the denominators are like (matching) in the same way that it matters when adding or subtracting fractions.

An example with numbers is shown below, along with a symbolic representation using *a, b, c,* and *d* to represent any numbers. Don't be intimated by the notation or its abstractness. It's just yet another way of reminding us to multiply straight across the numerators, and straight across the denominators.

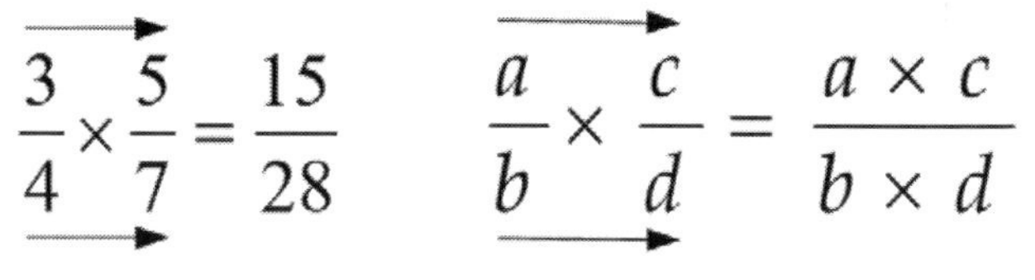

$$\frac{3}{4}\times\frac{5}{7}=\frac{15}{28} \qquad \frac{a}{b}\times\frac{c}{d}=\frac{a\times c}{b\times d}$$

Multiplying fractions "straight across"

Let's look at a somewhat practical example of multiplying fractions. Imagine you have a pizza pie, and you want to eat ½ of ½ of it. You could start by cutting it in half. Then you could take one of the halves, and cut away half of that half to eat. Think about what fraction of the whole pie you have eaten. You have eaten ¼ of it.

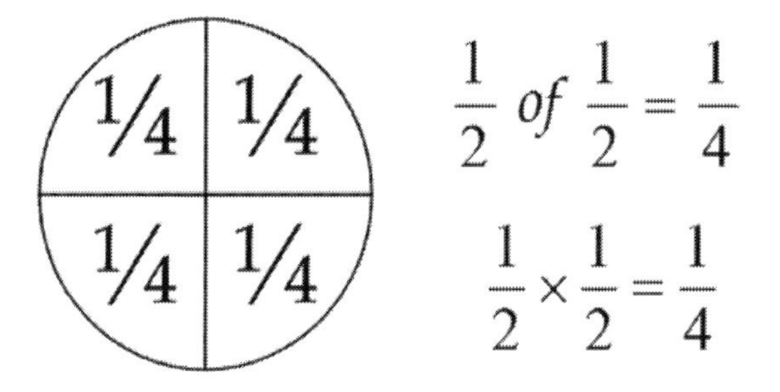

Computing ½ of ½ by multiplying "straight across"

When we use the word "of" in between two values, it means to multiply. What we really did was multiply ½ times ½ to get ¼. Remember that to multiply fractions, we multiply "straight across" both top and bottom.

WHAT IS A RECIPROCAL?

In non-mathematical terms, the **reciprocal** of a fraction is obtained by "flipping" the fraction "upside-down" so the numerator becomes the denominator, and the denominator becomes the numerator. For example, the reciprocal of 3/7 is 7/3. Understand that those fractions are *not* equal to each other. They are just reciprocals of one another.

$$\frac{3}{7} \neq \frac{7}{3}$$

There are circumstances in which we need to make use of the reciprocal of a fraction, so it's important to understand what it is, and how to form it. One such circumstance is described in the next section.

Note that the only time a fraction is equal to its reciprocal is when the numerator and the denominator are the same. For example, 3/3 = 3/3 even after you "flip" it.

DIVIDING A FRACTION BY A FRACTION

Dividing a fraction by a fraction is easy as long as you follow the simple, four-step procedure below, and don't confuse it with anything else. What we're going to do is change our fraction division problems into multiplication problems which we just learned how to do.

Step 1 of 4: **Leave** the first fraction alone
Step 2 of 4: **Change** the division to multiplication
Step 3 of 4: **"Flip"** the second fraction to its reciprocal
Step 4 of 4: **Multiply** as previously described

Look at next example. We left the first fraction alone. Then we changed the division operation to multiplication. Then we "flipped" the second fraction "upside-down" to its reciprocal. Then we multiplied the fractions.

$$\frac{3}{8} \div \frac{7}{11} = \frac{3}{8} \times \frac{11}{7} = \frac{3 \times 11}{8 \times 7} = \frac{33}{56}$$

The four-step procedure for dividing fractions

If it helps, you can just remember, "Leave, Change, Flip, Multiply," as long as you don't forget what each of those terms mean. Resist the temptation to not follow the steps, or to skip any, or to reorder them.

SO NOW WHAT?

Before progressing to the next chapter, it is essential that you fully understand all of the concepts in this one. The next chapter introduces more advanced fraction topics. If you don't fully understand this chapter, the next one will probably be confusing and difficult.

Be sure to also study the multiplication table in Chapter Two since that will play a large role in the upcoming material. See the Introduction and Chapter Twelve for information about how to get help, ask questions, or test your understanding.

CHAPTER SIX

More About Fractions

A FRACTION IS ACTUALLY A DIVISION PROBLEM

In the last chapter we defined a fraction as a value that represents part of a whole. While that is certainly true, there is another important way of thinking about fractions that must be fully understood. A fraction is actually a division problem in and of itself. More specifically, a fraction's value can be computed by dividing its numerator by its denominator. More informally, this can be thought of as "top divided by bottom." The horizontal line in a fraction is actually a division line.

THE FOUR WAYS TO REPRESENT DIVISION

It is important to recognize that there are actually four ways to represent division. We just learned that a fraction of the form "*a* over *b*" is another way of representing "***a*** divided by *b*." Look at the next example and make sure you understand what is being divided, and what is doing the dividing—it's "top divided by bottom."

$$a/b = a \div b = \frac{a}{b} = b\overline{)a}$$

Four equivalent ways of writing "*a* divided by *b*"

So we now have two different ways of thinking about a fraction. For example, you can think of the fraction 1/8 as representing one slice of a pizza pie which was cut into eight equal slices. You can also think about it as one whole pie divided into eight parts.

TREATING A FRACTION AS A DIVISION PROBLEM

Some of this will make more sense after studying the material in the Chapter Nine on decimals. In fact, you'll learn that to convert a fraction to a decimal number, you must treat the fraction as a division problem and compute "top divided by bottom," and not vice-versa.

For now, just make sure you understand that a fraction is a division problem in and of itself. However, don't confuse that fact with a problem that involves one fraction divided by another fraction like we saw in the previous chapter. When dividing a fraction by another fraction, just follow the four-step procedure described in the previous chapter.

CONVERTING AN INTEGER TO A FRACTION

There are many times when we need to work with an integer in fraction form. The very simple rule to remember is that we can convert an integer into an equivalent fraction by putting the integer over a denominator of 1.

It's important to understand why we are allowed to do this. Any number divided by 1 is equal to itself. Remember that a fraction is a division problem, so putting a number "over 1" means to divide it by 1. Dividing a number by 1 doesn't change it, so there isn't any harm in doing so. This is represented symbolically below.

$$n = \frac{n}{1}$$

Convert a number to a fraction by putting it "over 1"

MULTIPLYING AN INTEGER TIMES A FRACTION

Sometimes we need to multiply an integer times a fraction. Recall that the word "of" means multiplication when it occurs between two values. So, if we wanted to compute $\frac{1}{5}$ of 60, we would have to multiply $60 \times \frac{1}{5}$. The common *mistake* is to multiply both the numerator and the denominator of the fraction times the integer. That is *incorrect* for several reasons. Let's look at an example.

Imagine a pizza pie that was unusually cut into 7 equal slices, and imagine that you ordered and ate 3 of those slices. You can say that you ate 3/7 (three-sevenths) of the pie. Now imagine that you were still hungry after, and decided that next time you will order twice as much pizza. You'll have to multiply the 3/7 portion times 2.

As mentioned, the common *mistake* is to multiply both top and bottom by the integer. If you did that, you'd end up with 6/14. Think about what that means. Imagine

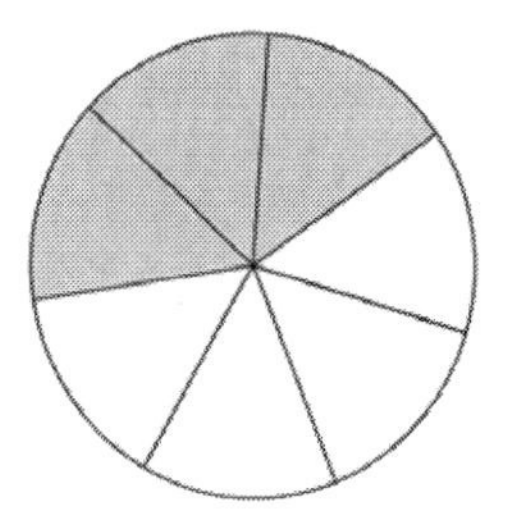

that the same pie of 7 slices was instead cut into 14 slices. It's still the same pie. It's still the same total amount of food, but cut into more and smaller parts. Make sure you see that if you ate 6 of those thinner slices, it's the exact same amount of food as if you ate 3 of the original-sized slices. It's not the double portion of food that you had intended to eat.

This will make a bit more sense later in this chapter when we discuss equivalent fractions, but make sure that you intuitively see that 3/7 = 6/14. We do *not* double the value of a fraction by doubling both the numerator and denominator. Instead, if we want to multiply a fraction times an integer, in this case 2, we only multiply the numerator by that integer. Let's take a look at why.

In practice, if we wanted to eat double the amount of food in the previous example, we would need to eat 6 slices of the original pie instead of 3 for a fraction of 6/7. We're not changing the way in which the pie is cut, we're just doubling the slices eaten. That implies that we must only multiply the numerator times the integer, and leave the denominator alone. The pie still has 7 total slices.

$$2 \times \frac{3}{7} = \frac{2}{1} \times \frac{3}{7} = \frac{6}{7}$$

$$a \times \frac{b}{c} = \frac{a}{1} \times \frac{b}{c} = \frac{a \times b}{c}$$

The integer only multiples the numerator

From a mathematical point of view, recall that we can convert an integer to a fraction by putting it over a denominator of 1. By doing so, we now have a problem involving a fraction times a fraction, which we've already learned how to solve. We just multiply straight across the top and bottom, giving us 6/7.

The simple rule to remember is that when multiplying an integer times a fraction, the integer only multiplies the numerator. This is because it is always over an "imaginary 1," even if we don't actually write it as such. Whenever you're doing problems in the world of fractions, get into the habit of always visualizing integers over a denominator of 1, even if you don't write it.

FRACTIONS WITHIN FRACTIONS

Take a look at the example below. We have a fraction comprised of fractions instead of integers like we've been working with. It looks far worse than it is. First, we need to locate the main fraction bar. It is always longer than any other fraction bar in the problem.

Remember that a fraction is just a value divided by a value. In this case the values in question happen to be fractions themselves. A fraction is just a special type of number. In this case we have (2/3) over (7/8).

$$\frac{\frac{2}{3}}{\frac{7}{8}} = \frac{2}{3} \div \frac{7}{8} = \frac{2}{3} \times \frac{8}{7} = \frac{2 \times 8}{3 \times 7} = \frac{16}{21}$$

A fraction comprised of fractions

Remember that a fraction can be thought of and rewritten as the numerator divided by the denominator, or more simply "top divided by bottom." In this case we happen to have a fraction on top and a fraction on the bottom, but that doesn't change anything. We can still rewrite the problem in the more conventional form.

Once we have done that, the problem is no different than what we worked with in the last chapter. It is simply a fraction divided by a fraction. Review the four-step procedure shown in the last chapter which is effectively "leave, change, flip, multiply."

This is is shown symbolically below, and is absolutely nothing different than what we learned in the last chapter. The task is simply to not get overwhelmed by the sight of the initial "fraction of fractions."

$$\frac{\frac{a}{b}}{\frac{c}{d}} = \frac{a}{b} \div \frac{c}{d} = \frac{a}{b} \times \frac{d}{c} = \frac{a \times d}{b \times c}$$

A symbolic representation of converting a fraction into a division problem, in this case comprised of fractions.

DIVIDING AN INTEGER BY A FRACTION

Now that we know that fractions within fractions are not a big deal, we can follow the same overall procedure to handle fractions in which there is a fraction in the numerator or denominator, but not both. All we need to remember is that an integer can always be written over a denominator of 1 if we are using it in a fraction problem.

In the next example, we have an integer divided by a fraction. The first step is to quickly rewrite the fraction in the more conventional "top divided by bottom" format so that we won't get confused. Then, recall that in order to work with the integer as part of a fraction problem, we must put it over a denominator of 1, and then proceed as we did in the previous example.

Again, the task is to simply not get overwhelmed, and not skip any steps. Just convert the problem into the more conventional division format, put the integer over 1 so that it functions a fraction, and then proceed with the four-step process to divide two fractions.

$$\frac{2}{\frac{3}{4}} = 2 \div \frac{3}{4} = \frac{2}{1} \div \frac{3}{4} = \frac{2}{1} \times \frac{4}{3} = \frac{2 \times 4}{1 \times 3} = \frac{8}{3}$$

A fraction comprised of an integer divided by a fraction

DIVIDING A FRACTION BY AN INTEGER

The next example is very similar to the one we just worked with, except the integer is in the denominator instead of the numerator. Again, just take it one step at a time. Don't skip any steps, and don't combine any steps.

Rewrite the fraction in the conventional "top divided by bottom format," remembering to convert the integer to a fraction by putting it over a denominator of 1. Then proceed as we've been doing.

$$\frac{\frac{2}{3}}{5} = \frac{2}{3} \div 5 = \frac{2}{3} \div \frac{5}{1} = \frac{2}{3} \times \frac{1}{5} = \frac{2 \times 1}{3 \times 5} = \frac{2}{15}$$

A fraction comprised of a fraction divided by an integer

FRACTIONS THAT ARE EQUAL TO 1

Remember that a fraction is a division problem whose value is found by computing "top divided by bottom." If a fraction's numerator and denominator are equal, the fraction is equal to 1. For example, the fraction 3/3 is equal to 1 because 3 divided by 3 is 1. Just remember that any fraction of the form n/n is equal to 1, where n is any value, including negatives. Remember that a negative divided by a negative equals a positive. The importance of all this will made clear later in this chapter.

$$\frac{n}{n} = 1$$

Any fraction of the form n/n is equal to 1

FINDING THE GREATEST COMMON FACTOR (GCF)

In Chapter Three we defined **factor** by example. The factors of 12 are 1, 2, 3, 4, 6, and 12. All of those numbers divide into 12 evenly, that is to say with no remainder.

A very common task in math is to compute what we call the **greatest common factor (GCF)** of two numbers. We use the GCF when we need to reduce (simplify) a fraction, discussed later in this chapter. Finding the GCF of two numbers is done in four steps:

Step 1: List all of the factors of the first number
Step 2: List all of the factors of the second number
Step 3: Take note of the factors that appear on both lists
Step 4: Choose the largest of those common factors

For example, let's find the GCF of 24 and 36. The factors of each number are listed below with the common factors bolded. It is easy to see that 12 is the greatest of the common factors.

Factors of 24: **1**, **2**, **3**, **4**, **6**, 8, **12**, 24
Factors of 36: **1**, **2**, **3**, **4**, **6**, 9, **12**, 18, 36

When thinking about factors, it is very important to be in the mindset of "smaller than." The factors of a number are all *less than* that number, with the exception of the factor which is equal to the number itself. Even though our task is usually to search for the greatest factor in common between two numbers, the factors of a number are never greater than the number itself. Take a moment to carefully reread this paragraph.

As another example, let's find the GCF of 15 and 23. The factors of each number are listed below with the common factors bolded. As you can see, 1 is the greatest of the common factors. Very often that will be the case, the significance of which will be explained later.

Factors of 15: **1**, 3, 5, 15
Factors of 23: **1**, 23

For practice, write down various pairs of numbers, and follow the given four steps to compute the GCF for each pair. It is very important to get to the point where you can do this very quickly and accurately, and least for numbers between 1 and 144.

This will be much easier if you have fully memorized your multiplication table, and are able to recite across each row from memory. Understand that factors are essentially based on multiples of numbers. For example, 3 is a factor of 24 because 24 is a multiple of 3, which is turn means that 24 is evenly divisible by 3. That's what is means for a number to be a factor of another number.

Note that there are many different procedures for finding the GCF of two numbers. Some of these methods are quite extravagant and unrealistic for a typical exam situation. Simply practice to get to the point where you can perform the given four steps quickly, and ultimately in your head. In most cases you will not be given large or complex numbers to work with.

REDUCING (SIMPLIFYING) FRACTIONS

The reason why we spent all this time learning how to compute the GCF of two numbers is because we must use the GCF when we do what is called **reducing** or **simplifying** a fraction to lowest terms. As we will see, both of those terms are actually not technically accurate, but nonetheless those are the terms we use.

You may know from experience that if you add two fractions and get an answer of 4/8, it would be "incorrect" to write your answer as such. We say that such a fraction is not reduced—it can be simplified. There is some number that we can divide into both 4 and 8 in an effort to further "break down" those numbers. In math terms, there exists some common factor of 4 and 8 that we can divide into both the numerator and the denominator. This will be made more clear by example:

Using our four step procedure, determine that the common factors of 4 and 8 are 1, 2, and 4. The greatest of those common factors (the GCF) is 4. That is the number that we will use to simplify our fraction of 4/8.

What we do is simply divide both the numerator and the denominator by the GCF. In this case, the numerator becomes 1, and the denominator becomes 2. We say that the fraction has been "reduced" or "simplified" to 1/2.

$$\frac{4}{8} = \frac{4 \div 4}{8 \div 4} = \frac{1}{2}$$

We are "allowed" to divide the top and bottom of a fraction by the same number since that doesn't change its value

There are some very important points to make. First, understand that we did not actually change the actual value of the original fraction, even though the words "reduce" or "simply" may imply so. You probably know from experience that if you eat 4 slices of a typically-cut pizza pie, you ate half the pie. It's still the same amount of food no matter how you look at it.

It's important to understand that by dividing both the top and bottom of the fraction by 4, we essentially worked with the fraction 4/4. Recall that any fraction of the form *n*/*n* is equal to 1. Stated yet another way, we can say that we divided the original fraction by 1 (in a disguised form), and we know that dividing a number by 1 doesn't actually change it. That is why we are "allowed" to do this procedure. We didn't alter value of the original fraction at all. We simply "pulled out" a common factor from both the numerator and the denominator by dividing both by that factor.

Recall from the previous chapter that if we only alter the numerator or the denominator of a fraction, we are in fact changing the value of the fraction. That is something that we are "not allowed" to do, unless it's for the sake of truly altering the original fraction which is rarely the intent. This is discussed in detail in the next section.

USING A FACTOR, BUT NOT THE GCF

What happens if we follow the procedure for reducing a fraction, but instead of using the GCF, we use some other common factor? Perhaps we used a smaller common factor because we had trouble finding the largest. The answer is that nothing bad happens, and we will still be able to reduce the fraction to lowest terms. It will just take one or more extra repetitions of the procedure. See the example below. Using a common factor of 2 was not enough to fully reduce the fraction. There still existed a common factor (again 2) that could be "pulled out" from both the numerator and the denominator.

$$\frac{4}{8}=\frac{4\div 2}{8\div 2}=\frac{2}{4} \quad \rightarrow \quad \frac{2}{4}=\frac{2\div 2}{4\div 2}=\frac{1}{2}$$

By not using the GCF it takes two steps instead of one to reduce this fraction

The reason why we always try to find and use the GCF is because by doing so, we can simplify a fraction in one step. Think about reducing the fraction 24/36. If we follow our procedure using the GCF of 12, then in one step we can reduce the fraction to 2/3. If we used a smaller common factor such as 2, it would take many repetitions in order to fully reduce the fraction. There is nothing at all wrong with that, though. We will still get the same reduced fraction—it will just take more steps and more time.

How do we know when a fraction is fully reduced? In math terms, the answer is when the GCF of the numerator and the denominator is 1. We could certainly divide both the numerator and the denominator by 1, but it would get us nowhere since we know that dividing a number by 1 doesn't actually change it. An example would be the fraction 15/23. The GCF of 15 and 23 is 1. In plain English, there is nothing for us to "pull out." The fraction is already as reduced as it is going to get.

IF WE MULTIPLY OR DIVIDE THE TOP OF A FRACTION, WE MUST DO THE SAME TO THE BOTTOM

It is worth repeating the important concept presented in this chapter since it comes up so frequently in math. To ensure that we do not change the value of a fraction, whatever we do to the top of a fraction we must always also do to the bottom, if we are referring to dividing or multiplying both top and bottom by a given number.

Sometimes we will divide both the top and bottom of a fraction by the same value, and sometimes we will multiply both the top and bottom of a fraction by the same value. Whatever you do to the top you must also do to the bottom, or you will end up changing the value of the original fraction.

Admittedly we had an example earlier in this chapter in which we doubled the value of a fraction by only multiplying the numerator by 2, but in that example our intent was specifically to alter (namely double) the value of the original fraction. Our intent was not to "convert" the fraction into an equivalent one.

It's important to stress that this section involves multiplying and dividing top and bottom by the same number. With those operations we won't be altering the value of the original fraction if we do the same thing to both top and the bottom, which is why it's "allowed."

However, this is not the case with addition and subtraction. If we add a given value to the numerator of a fraction, or subtract a given value from it, we are not allowed to simply do the same thing to the denominator. Even though we did the same thing to both top and bottom, we will have altered the value of the original fraction which is "not allowed."

As a simple example, let's take the fraction 2/3, and add 1 to both the numerator and denominator. We get 3/4 which you know from experience has a totally different value than 2/3. If we instead subtract 1 from both the numerator and denominator we get 1/2, which of course is also a different value than 2/3.

We are never allowed to take it upon ourselves to change the actual value of a fraction. The only way to ensure that we don't is by either dividing both top and bottom by the same number, or by multiplying both top and bottom by the same number. Of course there is no harm in adding or subtracting 0 to both top and bottom, but that accomplishes nothing.

LISTING THE MULTIPLES OF A NUMBER

In preparation for the topic of adding and subtracting fractions with unlike (non-matching) denominators, it is essential to review the concept of the multiples of a number which was introduced in Chapter Three when we worked with the multiplication table. Stated simply, as we read across a row of the table or down a column, we see the multiples of the number in the row or column header. Some examples are below:

Multiples of 2: 2, 4, 6, 8, 10, 12, 14, 16, 18, 20, 22, 24, 26, etc.
Multiples of 3: 3, 6, 9, 12, 15, 18, 21, 24, 27, 30, 33, 36, 39, etc.
Multiples of 7: 7, 14, 21, 28, 35, 42, 49, 56, 63, 70, 77, 84, etc.

Remember that all numbers have themselves as a multiple. That is because if you multiply a number times 1, you get back the original number. We usually only concern ourselves with positive multiples of a number, but it is certainly the case that you can multiply a number times 0 to get 0, and you can multiply a number times negative numbers to get negative multiples.

MULTIPLES VERSUS FACTORS

When thinking about multiples, it is very important to be in the mindset of "greater than." The multiples of a number are all *greater than* that number, with the exception of the multiple which is equal to the number itself. This is in contrast to when we worked with the factors of a number, and we were in the mindset of "smaller than" the number. The factors of a number are all *less than* that number, with the exception of the factor which is equal to the number itself. Reread this section carefully.

COMPUTING THE LEAST COMMON MULTIPLE / LOWEST COMMON DENOMINATOR (LCM / LCD)

We've worked with the concept of greatest common factor (GCF). Now we need to completely switch mindsets and learn about the concept of **least common multiple (LCM)**. When we need to add or subtract two fractions with unlike (non-matching denominators), we need to compute the LCM of the two unlike denominators. We will then "convert" each fraction so that each one has the LCM as its common denominator, thereby allowing us to add or subtract them as we've learned how to do. This is explained in the next section.

ADDING AND SUBTRACTING FRACTIONS WITH UNLIKE (NON-MATCHING) DENOMINATORS

In the last chapter we learned how to add and subtract fractions with like (matching) denominators. It was very easy because all we did was add or subtract the numerators, and leave the common denominator alone.

In this section we'll learn how to add and subtract fractions that do not have a common denominator, such as 3/4 + 1/6, depicted below. Some important points must be made. Remember that when we work with fractions, we almost must think in terms of some generic whole. Don't say, "What if my slices came from a larger-sized pie than yours did?"

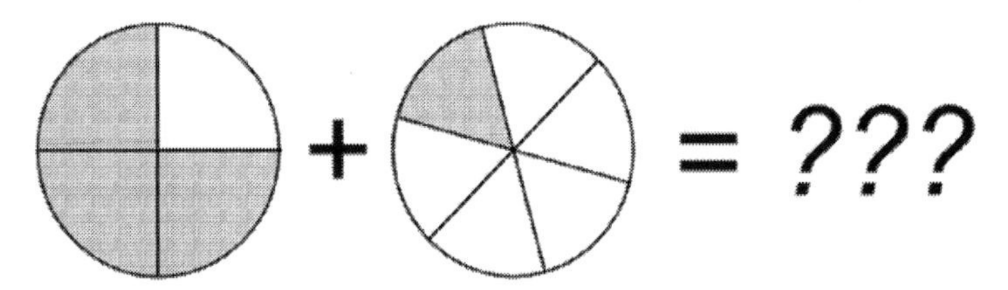

We can't directly add fourths and sixths since we're dealing with slices of different sizes

Make sure you see that we cannot just add the numerators and add the denominators. We're dealing with slices of different sizes since they came from differently-cut pies. It's not as simple as saying, "I ate one-eighth and you ate two-eighths, so combined we ate three-eighths."

In order to add the two fractions in the example, we somehow need to "convert" each one such that they have a common denominator. What we are going to do is compute the least common multiple (LCM) of the two denominators, and then we will use that as our **lowest common denominator (LCD)**.

LCM and LCD essentially mean the same thing for our purpose. We compute the LCM, and use it as the LCD. It is very important to understand that at no time will we actually change the value of either of the fractions, which is why the procedure is being referred to as "converting," in quotes. This is the four-step procedure:

Step 1: List the first few multiples of the 1st number
Step 2: List the first few multiples of the 2nd number
Step 3: Note the multiplies that appear on both lists
Step 4: Choose the smallest of the common multiples

If necessary, extend the lists in Steps 1 and 2 until you find the first common multiple.

For our example, we must find the LCM of 4 and 6, and use that as our LCD when we add.

Multiples of 4: 4, 8, **12**, 16, 20, **24**, 28, 32, 36, 40, 44, **48**, 56, etc.
Multiples of 6: 6, **12**, 18, **24**, 30, 36, 42, **48**, 54, 60, 66, 72, etc.

While those two numbers have many (indeed infinite) multiples in common, the lowest common multiple is 12. We will use 12 as our LCD. That is the "target" number for each denominator. Somehow we must "convert" each fraction to an equivalent fraction with a denominator of 12. We do that using this four-step procedure:

Step 1 of 4: Multiply the denominator of the first fraction by whatever number is necessary so it becomes the "target" denominator (the LCD).

Step 2 of 4: Multiply the numerator of the first fraction by same number that you used to multiply the denominator. Remember that whatever we do to the bottom, we must also do to the top.

Step 3 of 4: Repeat steps 1 and 2 for the second fraction. You will multiply top and bottom by a number other than the one used for the first fraction, but you must still end up with the same "target" denominator.

Step 4 of 4: You now have two fractions with like (matching) denominators. Add (or subtract) them as described in the previous chapter.

$$\frac{3}{4}+\frac{1}{6}=\frac{3\times\mathbf{3}}{4\times\mathbf{3}}+\frac{1\times\mathbf{2}}{6\times\mathbf{2}}=\frac{9}{12}+\frac{2}{12}=\frac{11}{12}$$

"Converting" the fractions so they each have the LCD

Some important points must be emphasized. We're not actually "unreducing" or "enlarging" or "unsimplifying" either fraction. We are "converting" each fraction into an equivalent fraction so that we'll have a common denominator to work with. We must never alter the actual value of any involved fraction.

Remember that what we are working with here are multiples of the denominators. This has nothing to do with factors. It doesn't matter that 2 happens to be a common factor of 4 and 6, and also happens to be the GCF. Remember that we use the GCF when reducing a fraction. When adding or subtracting fractions with unlike denominators, we use the LCM.

Understand that we must simply find some common multiple to use as a common denominator. Let's say we didn't see that 12 was the LCD, and we instead used 24 which is also a common multiple. It must be since 4 × 6 is 24. If you try it, you'll see that you end up with an answer of 22/24. That is actually the correct answer, but it needs to be simplified. We do that by dividing numerator and denominator by the GCF of 22 and 24 which is 2, giving us a reduced fraction of 11/12—the same answer we got by using the LCM. Using the LCM just allows us to avoid the unnecessary step of reducing at the end.

Remember: When we add or subtract fractions with unlike (non-matching) denominators, we do *not* just add the numerators and add the denominators. We must first "convert" each fraction so that the two fractions have a common denominator. Then we can add or subtract as we learned in the previous chapter.

THE DIFFERENCE BETWEEN GCF AND LCM

It's easy to get GCF and LCM confused. When we compute the GCF, we are looking for the *greatest* common factor which implies "bigger." However, remember that the factors of a number are always smaller than the number in question, with the exception of the factor which is the number itself. So, out of all the factors of two numbers, all of which are *less than or equal* to the numbers, we are looking for the *largest* one in common.

We typically use the GCF for the purposes of reducing or simplifying a fraction to lowest terms.

When we compute the LCM, the opposite is true. Even though we are looking for the *least* or *lowest* multiple in common which implies "smaller," the multiplies of a number are always bigger than that number, except for the one that is equal to it. So, out of all the multiples of two numbers, all of which are *greater than or equal* to the number, we are looking for the *lowest* one in common.

Take some time to carefully reread this section, reviewing the previous sections on GCF and LCD as needed.

SO NOW WHAT?

Before progressing to the next chapter, it is essential that you fully understand all of the concepts in this one. The next chapter introduces some more advanced fraction topics that are very important. If you don't fully understand this chapter, the next chapter will probably be very confusing and difficult for you.

Take time to review the material. See the Introduction and Chapter Twelve for information about how to get help, ask questions, or test your understanding.

CHAPTER SEVEN

Other Topics in Fractions

INTRODUCING MIXED NUMBERS

A **mixed number** is the sum of a whole number (an integer) and a fraction. We see them fairly often in everyday life, for example $5\frac{1}{2}$. The most important thing to understand about that notation is that there is an implied *addition* symbol between the integer and the fraction. We read the implied addition sign as "and."

$$5\frac{1}{2} = 5 + \frac{1}{2}$$

"Five and a half" means five wholes plus one-half of a whole

Look at the next example. The shaded portion is a representation of the mixed number $2\frac{1}{3}$. Think of it as two whole pizza pies, plus one-third of another pizza pie. Everything you have learned about fraction problems applies. We're talking about wholes in a general sense. There is no concept of one pie being larger than another. It is also implied that any slicing is done in equal parts.

Notice that the first two pies have been cut into thirds, but that doesn't even matter. We are still symbolizing two wholes plus one-third of an additional whole.

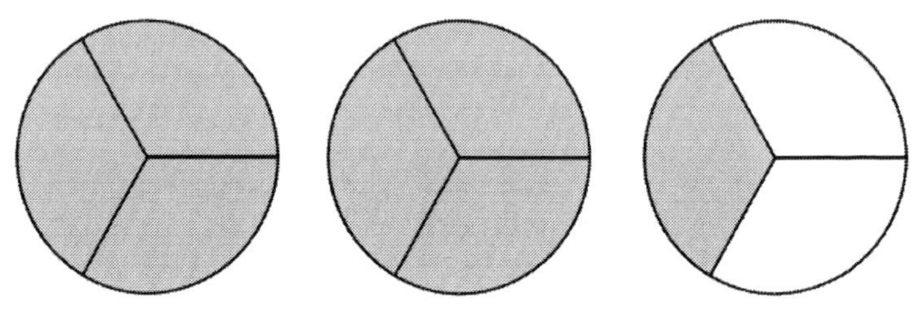

A representation of $2\frac{1}{3}$ (two and one-third)

INTRODUCING IMPROPER FRACTIONS

If the numerator of a fraction is equal to or larger than the denominator, we say that the fraction is an **improper fraction**. Some examples are 3/3, 5/1, and 7/2. Improper fractions come up frequently in math, and are discussed in the sections that follow.

CONVERTING MIXED NUMBERS TO IMPROPER FRACTIONS

While there are many everyday circumstances for which using mixed numbers is appropriate, when we have to do mathematical operations involving mixed numbers it is usually best to quickly convert them into improper fractions. If nothing else, we will then be dealing with a single fraction as opposed to the sum of a integer and a fraction which is more cumbersome.

The five-step procedure for converting a mixed number to an improper fraction is more complicated to explain than it is to do. As long as you understand and remember what a mixed number really is, you will not have any difficulty. Let's convert $2\frac{1}{3}$ to an improper fraction.

Step 1 of 5: Put a plus sign between the integer and fractional components of the mixed number

Step 2 of 5: Convert the integer to a fraction by putting it over a denominator of 1

Step 3 of 5: Note that the LCM of the two denominators involved will simply be the denominator of the second fraction, since the first fraction has a denominator of 1

Step 4 of 5: Multiply top and bottom of the first fraction by the LCM (i.e., the denominator of the second fraction) so that the fractions will have a common denominator.

Step 5 of 5: Add the fractions as previously described

$$2+\frac{1}{3}=\frac{2}{1}+\frac{1}{3}=\frac{2\times\mathbf{3}}{1\times\mathbf{3}}+\frac{1}{3}=\frac{6}{3}+\frac{1}{3}=\frac{7}{3}$$

Converting from a mixed number to an improper fraction

We found that 2 ⅓ is equivalent to 7/3. Look back at the pie diagram and notice how each of the whole pies could also be thought of as 3/3, or three slices out of three. The two whole pies make 6 slices which is actually 6/3, as strange as that may appear. We then have 1 additional slice from another pie, giving us the final 1/3. We could think of 7/3 as seven slices from pies which have each been cut into thirds.

There is a "shortcut" for the procedure which is worth exploring. We know that the integer ends up being multiplied by the denominator of the fraction, so we can just go ahead and do that. That product ends up being the numerator of the first fraction, which ends up getting added to the numerator of the second fraction, so we can go ahead and do that as well. The first fraction will be converted to have a denominator of 1, which is then multiplied by the denominator of the second fraction. All of that is represented symbolically below. It is good practice to make sure you understand the formula.

$$a + \frac{b}{c} = \frac{(a \times c) + b}{c}$$

Symbolically converting a mixed number to an improper fraction

CONVERTING IMPROPER FRACTIONS INTO MIXED NUMBERS

To convert an improper fraction into a mixed number, we will essentially perform the five-step procedure in reverse, but again, there is a shortcut.

Let's convert 7/3 back into a mixed number. One way to think about it is how we could go about reassembling the original pies from which the slices came. Three slices could be reassembled to form a whole pie, remembering that each pie started out comprised of three slices. Another three slices could make a second pie. We then have one slice left over, which we know is 1/3 of a whole pie. We have two whole pies plus 1/3 of a pie, which we can represent as 2 ⅓, which is sometimes typeset as 2 1/3.

We can also do our conversion mathematically. Remember that a fraction is simply a division problem—"top divided by bottom." In this case we can compute 7 ÷ 3, and we will have our answer. We know that 3 goes into 7 twice, and we have a remainder of 1. In earlier math we would represent that answer as 2 R 1. However, now that we're working with fractions, we can actually turn that remainder into a fraction.

The way it works is that we take the remainder, and put it over the denominator of the improper fraction that we are converting. In this case, we are converting 7/3, so we'll take the remainder of 1 and put it over 3, giving us a mixed number of 2 ⅓, as expected. This is described in detail the next section.

CONVERTING DIVISION PROBLEM REMAINDERS INTO FRACTIONS

Think about how you could divide 13 apples equally among three people. Each person could get 4 whole apples, and we would have one left over. How could we divide and distribute that apple fairly? Since it will be shared among three people, we could divide it into thirds. Each person would get 1/3 of the apple. We can say that 13 ÷ 3 = 4 ⅓. This is represented below.

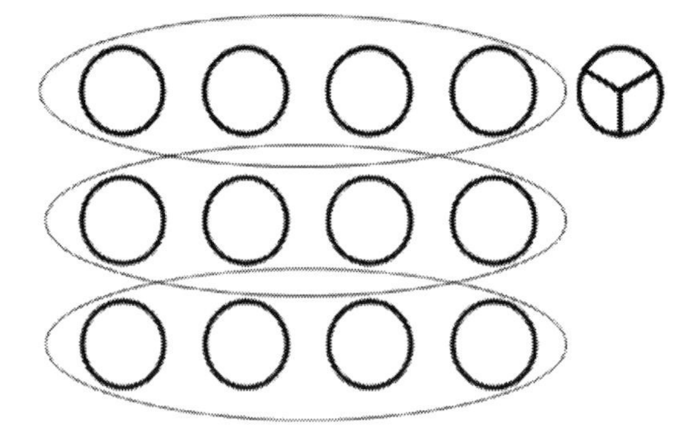

13/3 = 13 ÷ 3 = 4 R 1 = 4 ⅓

The rule to remember is that instead of leaving a remainder after we divide, we can represent our answer as a mixed number by taking the remainder and placing it over the original denominator in the problem, or equivalently, over that which you were dividing by.

OPERATIONS WITH MIXED NUMBERS

Sometimes we're asked to perform an operation involving mixed numbers such as $9\ \frac{2}{5} - 5\ \frac{3}{4}$. There are several different ways to solve such problems, all of which will result in the same answer. The most straightforward method is to simply convert all of the mixed numbers into improper fractions. That only takes a moment, and once that is done you can follow the steps you learned for performing standard operations involving fractions.

WORKING WITH EQUIVALENT FRACTIONS

We already saw that if we multiply both the numerator and the denominator of a fraction by the same value, the value of the fraction isn't altered. For example, we can multiply both top and bottom of a fraction by the same value to "convert" it for the purposes of adding or subtracting. We can also divide both top and bottom of a fraction by the same value in order to simplify it.

What we are doing in each of those cases is working with equivalent fractions—fractions that may look different, but actually have the same value.

Let's look at the two equivalent fractions 4/10 and 6/15. One way to know that these fractions are equivalent is to reduce them to lowest terms as we learned how to do. To reduce the first fraction we will use a GCF of 2, and to reduce the second fraction we will use a GCF of 3. In both cases, we end up with reduced fractions of 2/5. That proves that 4/10 and 6/15 are equivalent—they represent the same portion of a whole.

Another way to know that these fractions are equivalent is to compare the **cross products**. In plain English, a cross product is simply what you get when you multiply "diagonally." One cross product in this example is 4 × 15 which is 60. The other cross product is 10 × 6 which is also 60. The fact that the cross products are equal proves that the two fractions involved are equivalent. Fractions that are not equivalent will have unequal cross products.

$$\frac{4}{10} = \frac{6}{15}$$

Both of these equivalent fractions can be reduced to 2/5. Also, the cross products of 15 × 4 and 10 × 6 are equal.

The concept of cross products is represented symbolically below. Make sure you understand that we're not actually doing any operation with the two involved fractions other than a simple comparison. What we are doing with the cross products is essentially a "test." We are "checking" the cross products to see if they are equal to one another, and if they are, it proves that the two involved fractions are equivalent. It doesn't even matter what the number the cross products actually work out to be. All that matters is whether or not they are equal.

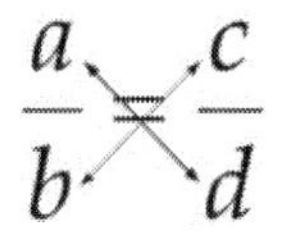

These two fractions are equivalent if ($a \times d$) = ($b \times c$)

CROSS MULTIPLYING VS. MULTIPLYING ACROSS

At this point in the material, many students get very confused with the difference between "cross multiplying" and "multiplying across" which you will recall is the procedure for multiplying two fractions.

When we multiply two fractions, we multiply "straight across" top and bottom. We multiply the numerators to get the numerator of the product, and we multiply the denominators to get the denominator of the product. We are performing the mathematical operation of multiplication on the two involved fractions.

That is different than cross multiplying. As described, cross multiplying is simply a test that we can perform whenever we are asked to determine whether two fractions are equivalent to each other. If the cross products are equal than the two fractions are equivalent.

For now, just make sure you don't forget the procedure for multiplying two fractions, and make sure you understand the concept of cross multiplying. It will be revisited later in the book when we discuss **proportions** which is the study of equivalent fractions.

"CROSS CANCELING" BEFORE MULTIPLYING FRACTIONS TO AVOID REDUCING THE PRODUCT

There is one last bit of confusing material on fractions which involves the word "cross." This topic is included because most students are drawn towards it, but it often leads to a great deal of confusion and errors.

Before multiplying two fractions, we can do what is informally referred to as "cross canceling" in an effort to simplify the arithmetic. We can optionally "pull out" a common factor from the tops and bottoms of the two involved fractions in an effort to not have to later reduce the answer after multiplying.

Look at the first example below. We know that when we multiply two fractions, we multiply across the top and then across the bottom. With that in mind, we can take note of the fact that we have a 3 somewhere on the top, and another 3 somewhere on the bottom. They aren't directly on top of one another, but that doesn't matter since multiplication is commutative. One 3 might as well have been on top of the other.

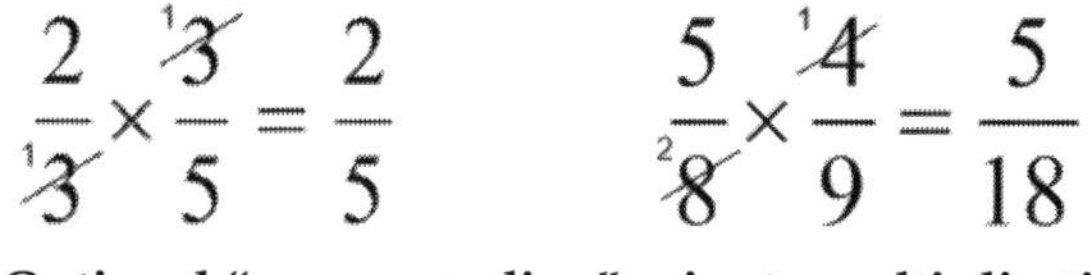

Optional "cross canceling" prior to multiplication

We can informally say that the threes "cancel out." If we just multiply as usual we will end up with 6/15. Since we can divide top and bottom by a GCF of 3 in order to reduce the fraction to 2/5, we might as well have just "pulled out" the threes in the first place. It is important to understand that the threes didn't just "disappear." In essence we were dealing with 3/3, which we know is equal to 1. Each 3 becomes a 1 as shown.

In the second example, note that the 4 on the top and the 8 on the bottom share a common factor of 4. Again, it doesn't matter that the 4 and the 8 aren't directly on top of each other. We can divide the top 4 by 4, making it a 1, and divide the bottom 8 by 4, making it a 2. This is permitted since it's no different than if we multiplied the original fractions and then divided both the numerator and denominator of the product by the GCF of 4.

It is important to *not* utilize this shortcut when dividing a fraction by another fraction. Recall that to divide two fractions, we must do several steps to convert the problem into multiplication. Do *not* do any type of "cross canceling" prior to that step, no matter how tempting the numbers might make it seem.

It is also important to not confuse the "cross canceling" shortcut with cross multiplying. Recall that we can use cross multiplying to determine if two fractions are equivalent. That is a totally separate procedure from the one described in this section.

NEGATIVE FRACTIONS

Negative fractions are a source of confusion for many students, and they play a large role in later math. For now it's important to understand the general concept. We said that we can think of negative values as how much we owe. Imagine you ate three-fourths of your friend's pizza pie, and now you owe him/her that amount. There are three ways we can write ¾ as a negative value. We can put a negative sign in the numerator, or in the denominator, but not both. We can also just write the fraction as a positive, and then put a negative sign to the left to make it negative.

$$\frac{-3}{4} = \frac{3}{-4} = -\frac{3}{4}$$

Three equivalent ways of writing "negative three-fourths"

It is important to understand why this makes sense. Remember that a fraction is a division problem—top divided by bottom. Also remember that a when we divide, we will get a negative quotient if we divide numbers of opposite signs. A negative divided by a positive is negative, and a positive divided by a negative is negative. That takes care of the first two versions shown above. We can also just take the entire positive value of ¾ and make it negative by putting a negative sign to the left of it. It is no different than taking a positive whole number such as 7 and making it negative by writing a negative sign to the left of it.

Don't get confused and write a negative fraction with a negative sign in both the numerator and the denominator. Recall that when we divide values with matching signs, we get a positive quotient. A negative number divided by a negative is positive. The negative signs "cancel out," in a sense, leaving us with a positive fraction. Be sure that you understand this concept.

$$\frac{-3}{-4} = \frac{3}{4} \qquad \frac{-3}{-4} \neq -\frac{3}{4}$$

The two negatives "cancel" to make a positive fraction

0 DIVIDED BY ANYTHING (OTHER THAN 0) IS 0

It's easy to get when confused when doing division problems involving 0. Rather than memorizing the rules, it is better to just think about what you are trying to compute. Let's think about the problem 0 ÷ 17. Remember, that is "zero divided by 17." The scenario is that you have zero apples, and you want to divide and distribute them evenly among 17 people. How many apples will each person get? The answer is 0. There are no apples to go around. We can write this rule symbolically as shown below, where n is used to represent any number.

$$0 \div n = 0$$

0 divided by any number (other than 0) equals 0

ANYTHING DIVIDED BY 0 IS UNDEFINED

Now let's examine the situation in which it is some number that is being divided by 0. A non-mathematical way of thinking of the situation is as follows: Imagine you have 100 apples, and you're trying to divide and distribute them evenly them among a certain number of people. In some cases the apples would divide evenly, such among 25 people, or 10 people, or 2 people. In some cases you could divide them as evenly as possible, and then you would have a few left over which you would have to cut up into pieces, but somehow or other you'd be able to distribute all of your apples.

Now think about what would happen if you wanted to divide the 100 apples among 0 people (not even yourself). It's not possible. All would be left since there is no one to give them to. In math, we define such as situation as being **undefined**. The answer is not 0, the answer is "undefined." We are "not allowed" to divide by 0. If you try it on a calculator, you will get an error message.

$$n \div 0 = Undefined$$

Any number divided by 0 is undefined

To review, we are allowed to divide 0 by a number. The answer is 0. Think of it as, "Everyone gets 0 apples." However, we cannot divide a number by 0. That is undefined. Think of it as, "I don't have anyone or even myself to share my apples with."

SO NOW WHAT?

Before progressing to the next chapter, it is absolutely essential that you fully understand all of the concepts in this one. The next chapter introduces some new topics involving fractions. If you don't fully understand this chapter, the next chapter will probably be confusing and difficult for you.

Take time to review the material. See the Introduction and Chapter Twelve for information about how to get help, ask questions, or test your understanding.

CHAPTER EIGHT

The Metric System, Unit Conversion, Proportions, Rates, Ratios, Scale

WHAT IS THE METRIC SYSTEM?

Virtually every country other than the United States uses the Metric System for everyday measurement. It is also used by the scientific community in the US. The Metric System is ideal for working with measurements because it is based on groups of 10, just like our math system.

Many test questions are designed to see if you feel comfortable working within the Metric System. The topic just requires some basic memorization and practice.

COMMON UNITS OF MEASURE IN THE CUSTOMARY OR IMPERIAL SYSTEM

Before introducing the Metric System, it's worth reviewing some common units of measure in the non-Metric

System which is usually referred to as the imperial or customary system. Before taking any standardized exam, determine whether or not you are expected to memorize these units of measure, or if they will be provided for you on a reference sheet.

The basic unit of length is the inch. There are 12 inches in a foot, and there are 3 feet in a yard. You might also be expected to know that there are 5280 feet in a mile.

The basic unit of mass (weight) is the ounce. There are 16 ounces in a pound, and there are 2000 pounds in a ton.

The basic unit of volume (capacity) is the cup. A cup is defined as 8 fluid ounces, but the actual scale weight of a filled cup depends upon what substance it is filled with. There are two cups in a pint, and there are two pints in a quart. There are four quarts in a gallon.

PROBLEMS INVOLVING TIME SPANS

Some test questions involve aspects of clock time. Make sure you know that there are 60 seconds in a minute, and 60 minutes in an hour. There are 24 hours in a day, and 7 days in a week. There are 12 months in a year. A non-leap year has 365 days and a leap year has 366 days.

It is important to memorize some basic fractions of time that come up very frequently. Since there are 60 seconds in a minute, and 60 minutes in an hour, the fractions will all involve a denominator of 60, and will be reducible. One-quarter (one-fourth) of 60 is 15. One-half of 60 is 30. One-fifth of 60 is 12, and one-sixth of 60 is 10. With this information we can deduce, for example, that 12 minutes is one-fifth of an hour (12/60 = 1/5), and so forth.

MEASURING TEMPERATURE IN BOTH SYSTEMS

In the Metric system, temperature is measured in degrees Celsius (°C). The system was designed so that water freezes at 0 °C and boils at 100 °C. A typical room temperature is about 21 °C. The non-Metric system measures temperature in degrees Fahrenheit (°F). Water freezes at 32 °F and boils at 212 °F. A typical room temperature is about 70 °F.

In later math you will learn the algebraic formulas for converting temperatures from one system to another. Until then, don't get flustered if you see a problem that happens to be based on degrees Celsius. Just work with the numbers no differently than you would if they were in the Fahrenheit system.

COMMON UNITS OF MEASURE IN THE METRIC SYSTEM

The Metric System is actually very straightforward. The basic unit of length is the meter. It is roughly equal to a yard. The basic unit of volume (capacity) is the liter. It is roughly equal to a quart. The basic unit of mass (weight) is the gram. A gram is extremely light. You would need almost 30 grams just to equal one ounce, so heavier items are more practically weighed in kilograms, which are units of 1000 grams. This is discussed in the next section.

COMMON METRIC SYSTEM PREFIXES

We noted that the metric system uses the meter, gram, and liter as its basic units. To represent larger or smaller quantities of those units, we work with special prefixes which tell us how many times we must multiply or divide the unit by a power of 10.

For example, the prefix "kilo-" tells us to multiply the unit times 1000. We can apply the prefix to any of the three basic units. For example, a kilogram is 1000 grams, which is equal to about 2.2 pounds.

The prefix "milli-" tells us to divide the unit by 1000, making it much smaller. For example, a milliliter is

1/1000 (one-thousandth) of a liter. It would take about 20 drops of water to fill the volume of a milliliter.

The prefix "centi-" tells us to divide the unit by 100, making it 100 times smaller, but 10 times larger than what the prefix "milli-" would do to it. For example, a centimeter is 1/100 (one-hundredth) of a meter. You've probably worked with rulers with one side divided into centimeters. A centimeter is a bit less than half an inch.

While there are other prefixes that represent other powers of 10 that can be applied to the basic units, by far the most common ones are "kilo-," "milli-," and "centi-." Be sure to remember what they mean. Later in this chapter we will learn how to convert units within the Metric system.

INTRODUCTION TO RATIO AND RATE

A **ratio** is a comparison of two quantities which have the same units. For example, if we compare the number of boys to girls in a class, the common unit is "students" or "people." If we compare the amount of sugar needed for a recipe to the amount of flour, the common unit would be some measure of volume such as cups.

We do our comparison using division, which means that a ratio is effectively a fraction. For example, a baker may know that in a certain recipe, for every 3 cups of sugar used, s/he must use 5 cups of flour. We could say that the ratio of sugar to flour is 3 to 5. We are effectively dealing with the fraction 3/5, and we can work with that fraction no differently than any other. This is discussed later in the chapter in the section on proportions.

$$\frac{3\ C\ sugar}{5\ C\ flour}$$

THREE DIFFERENT WAYS OF WRITING A RATIO

There are three different ways of writing a ratio, all of which are equivalent. To express the ratio of *a* to *b*, we could write precisely that using the word "to." We could also write *a:b* using a colon, or we could write $\frac{a}{b}$ using fractional representation. The order in which we write the two numbers is significant. They are not interchangeable, and must match the order in which we describe what is being compared.

$$a \text{ to } b \qquad a{:}\,b \qquad \frac{a}{b}$$

Three different ways of representing "the ratio of *a* to *b*"

INTRODUCTION TO RATES

$$\frac{7\ cm.}{2\ min.}$$

A **rate** is basically the same as a ratio, but it involves the comparison of two quantities that have different units. For example, if a snail travels 7 centimeters in two minutes, we are comparing a unit of distance to a unit of time. We will still work with a fraction of 7/2 to match the order of the values that we are referring to, but we must include the units in our rate, and present it as shown at left. When describing a rate, we typically use the word "per" as in "30 miles per hour." Even though this would be written in everyday life as 30 mph, when we work with such a rate in math we will write it in fraction form as *30 mi. / 1 hr.*

WORKING WITH UNIT RATIOS AND RATES

We often "reduce" ratios and rates so that their denominator is 1. We use the word "unit" to describe such ratios and rates since they are based on 1 single unit of whatever is being compared. For example, while it is true that 36 inches equals 3 feet, we would usually express this relationship as the reduced *12 in. / 1 ft,* obtained by dividing both of the involved numbers by 3. If apples were priced at $7 for 2 pounds, we would likely express this as the reduced *$3.50 / 1 pound* ($3.50 per pound), obtained by dividing both of the involved numbers by 2.

$$\frac{12\ in.}{1\ ft.} \qquad \frac{\$3.50}{1\ pound}$$

Two examples of unit rates

COMPUTING THE COST PER UNIT

The most common problem on the topic of unit rates is to compute the cost for one unit of something when we are provided with the total cost of many such items. For example, a problem may state that a person is selling a box of 17 apples for $13.95, and ask us to compute the cost of just 1 apple at that rate.

To solve this problem we must arrange the given values into a fraction, but we must determine which must go on the top and which must go on the bottom. It will not necessarily match the order in which they were given. In this case we are interested in "cost per apple," so we'll need to put the dollar amount on the top and the number of apples on the bottom. This is almost always what we will do when solving rate problems involving money.

We now have *$13.95 / 17 apples,* but we want the denominator to be 1 since we want to know the cost for just 1 apple. We can make the denominator 1 by dividing it by 17, which we must remember to do to the numerator as well so that we end up with an equivalent rate. We end up with *$0.82 / 1 apple* (82 cents per apple), which is our answer rounded to the nearest cent.

$$\frac{\$13.95\ (cost)\ \div 17}{17\ (apples) \div 17} = \frac{\$0.82}{1}$$

Converting a rate to an equivalent unit rate

Note that in essence all we did was divide the numerator by the denominator, namely $13.95 ÷ 17, but it is important to understand the process behind it since it is very easy to get confused and do things wrong, or in the wrong order.

CONVERTING MEASUREMENTS WITH UNIT RATIOS

Unit ratios make it easy for us to convert measurements from one unit to another. For example, we may be asked to compute how many feet are equal to 48 inches. Most students immediately know that they have to do something involving the number 12 since there are 12 inches in a foot, and can deduce that they must divide by 12 as opposed to multiplying by it.

However, in some situations it can be confusing as to whether or not we must divide or multiply by the number that we know will be involved. We can solve such problems by using a **unit ratio** which has a denominator of 1. The general idea is that we will multiply the given value by a unit ratio such that the unit we are converting from will be "canceled out," and we will be left with the unit to which we want to convert.

Let's try converting 48 inches to feet using a unit ratio. The first thing we must do is put the 48 inches over a denominator of 1 in order to convert it into a fraction. Recall that we are always allowed to do that. We know that there are 12 inches in 1 foot, and as part of our problem we're going to use the unit ratio *12 in. / 1 ft*, or its reciprocal. Recall that a ratio is really just a fraction.

Since 12 inches equals 1 foot, the fraction *12 in. / 1 ft.* actually equals the number 1, since the numerator and the denominator are really the same. That is also true of the reciprocal *1 ft. / 12 in.* if it turns out that we need to use that instead. It's no different than the fraction 7/7 being equal to 1 which we learned. We also learned that there is never any harm in multiplying a value times 1, since doing so doesn't change it.

What we're going to do is multiply our fraction of *48 in. / 1* times our unit ratio such that the units of inches "cancel out," and the units of feet remain, since that is what we want to convert to. We need to set up our unit ratio so that inches are on the bottom and feet are on the top. This way, the inches that are on the top can "cancel" with the inches that are on the bottom. The term "cancel" is actually very non-mathematical, but most students feel comfortable with the concept, and it will serve our purpose. We must set up our problem as shown below.

$$\frac{48\ \cancel{inches}}{1} \times \frac{1\ foot}{12\ \cancel{inches}} = \frac{48}{12}\ feet = 4\ feet$$

Converting a measurement's units using a unit ratio

Before we multiply straight across like we do when multiplying any two fractions, we can simply cross off the unit of inches that are on the top of the first fraction and the unit of inches that are on the bottom of the second fraction. The only unit that remains is feet which is what we wanted. Recall that a fraction is really just a division problem, so our computation is "top divided by bottom" or 48 ÷ 12. In some cases the numbers will "work out" nicely like they did in this problem. In other problems you may need to reduce the final fraction using the techniques previously described. In some cases it will make sense to convert the final fraction into a decimal which you'll learn how to do in the next chapter.

Understand that if we had used the reciprocal of our unit ratio, namely *12 in. / 1 ft.*, the units of inches would not have canceled, and the number we would have ended up with would have been totally wrong.

Let's try one more example and convert 500 milliliters (mL) to liters (L). For problems involving the Metric system it extremely easy to get confused and not know if you should divide or multiply by a given value.

We'll start by putting 500 mL over a denominator of 1. We must then multiply that fraction times a unit ratio which has milliliters on the bottom and liters on the top since we want milliliters to cancel, and liters to remain. A milliliter is one-thousand*th* of a liter, so that means that there are 1000 milliliters in a liter. We'll put 1 liter on the top, and 1000 milliliters on the bottom as shown next.

$$\frac{500\ \cancel{mL}}{1} \times \frac{1\ L}{1000\ \cancel{mL}} = \frac{500}{1000}\ L = \frac{1}{2}\ L$$

Converting a measurement's Metric units using a unit ratio

Following the same procedure that we used for the last example, we end up with an answer of ½ L, obtained by reducing the fraction 500/1000. We've found that that 500 mL is equal to one-half of a liter.

It is very easy to get confused with problems like this, which is why it is important to take the extra time needed to carefully follow these steps. Most errors with problems like this are due to rushing and guessing.

INTRODUCTION TO PROPORTIONS

A **proportion** is simply a way of showing that two ratios (which are effectively fractions) are equivalent. For example, you probably feel comfortable with the fact that 1/2 and 4/8 are equivalent. We get the second fraction by multiplying both top and bottom of the first fraction by 4, which we learned is OK because it doesn't change the actual value of the first fraction. We can say that one-half and fourth-eighths are in proportion, and we can show this by way of a simple equals sign.

Many proportion problems involve solving for an unknown value. We are asked to determine the value that will make the proportion true. Look at the first example below. Notice that the denominator of the second fraction is 7 times as large as the denominator of the first fraction. Recall that we are permitted to multiply the denominator of the first fraction by 7 in order to obtain an equivalent fraction, but in the process we must also multiply the numerator by 7 so that we don't actually change the original fraction's value. Multiplying the numerator by 7 gives us 14 as the answer to the unknown value. We can say that 2/3 and 14/21 are "in proportion," or are proportional.

$$\frac{2}{3}=\frac{?}{21} \qquad \frac{1}{4}=\frac{5}{?}$$

Some sample problems involving proportions

In the second example we could first take note of the fact that the second fraction's numerator is 5 times as large as the first fraction's numerator, and therefore the second fraction's denominator must be 5 times as large as well, namely 20.

There is another way to look at the same problem. We could note that the denominator of the first fraction is 4 times as large as its numerator, and therefore the denominator of the second fraction will have to be 4 times as large as its numerator as well. As expected, we still get 20 if we think about the problem in that way.

Many proportion problems come in the form of word problems, or problems involving a diagram which we will work with later in this chapter. All that is necessary for such problems is to read or look at them very carefully, and convert them into a proportion. You'll need to determine exactly what piece of information is missing. You'll also have to ensure that if you comparing "apples to oranges" on the left, you're doing the same comparison in the same order on the right.

A COMMON MISTAKE INVOLVING PROPORTIONS

$$\frac{5}{6} \neq \frac{7}{8}$$

Many students wrongly believe that 5/6 is proportional to 7/8, with the logic that the left-side denominator is one more than the numerator, and the same is true of the fraction on the right. The student may also note that the numerator on the right-side fraction is two more than the numerator of the fraction on the left, as is true for the denominators.

This is not how proportions work. Any comparisons that we do between numbers in a proportion must be done by way of multiplication or division, and not by addition or subtraction. For example, in our previous problems we noted that some values were two or three *times* as large as others. That implies multiplication, and not addition. We say that 8/12 is proportional to 4/6 because the figures on the right are half of the figures on the left. That implies division (by 2 in this case), but doesn't imply subtraction.

$$\frac{8}{12} = \frac{4}{6}$$

INTRODUCTION TO SCALE DRAWINGS

A **scale drawing** problem is simply a proportion problem in picture format. A map is an example of a scale drawing. To determine the actual distance between two points, you could measure it on the map using a ruler, and then refer to the map's **scale** which compares a unit of distance on the map to what that unit actually represents. The map's scale is simply a ratio.

Some scale drawings are representations of three-dimensional objects. For example, in the next figure, a two-dimensional flat drawing of a bike is being used to represent what we know is a real three-dimensional bike.

Even if a drawing is designed to try to illustrate some sense of depth, it doesn't change the problem at all.

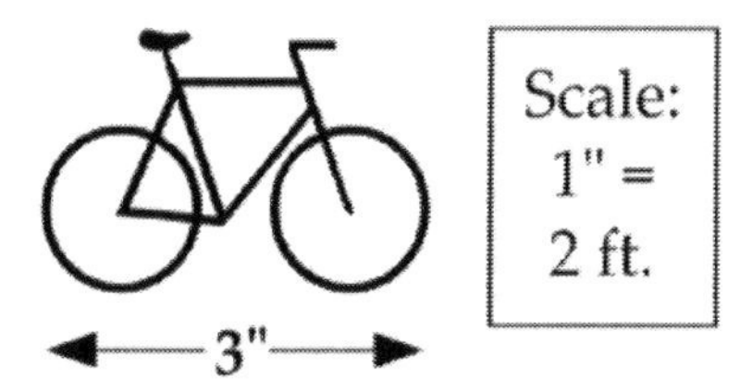

A scale drawing and legend for a problem which might ask us to compute the length of the actual bicycle.

All we must do with a scale drawing is accurately convert it into a proportion problem. Once that is done, the problem becomes no different than the ones discussed in the last section. We will almost always be given three out of four pieces information, and we'll have to find the missing piece just like we've done.

For the scale drawing in our example, we can see in the accompanying **legend** (scale description) that one inch in the drawing represents two feet in reality. We can immediately set up one side of our proportion. We'll put the inches on top and the feet on the bottom, but we could have actually done the opposite. All that matters is that both sides of the proportion match each other. If we're going to have inches on top on one side, we will need to have inches on top on the other.

The picture implies that the drawing of the bicycle is three inches long. It is very important to understand that many scale drawings are what we call "not drawn to scale." As confusing as that may sound, all that means is that we should use the numbers we are given, rather than attempt to do any measuring on our own. For example, the bicycle drawing appears to be quite a bit smaller than three inches, but we will work with it as though it is three inches. Only in the very lower grades would you be expected to take out a ruler and measure a drawing on your own.

To continue preparing our proportion, we must arrange the right side so that inches are on the top and feet are on the bottom. We're going to work with three inches for the drawing of the bike. We don't know how many feet the actual bicycle is, and that is almost certainly what such a problem will ask us to determine. We'll have to indicate that we don't have that information, and for now, we can just use a question mark for that.

$$\frac{1\ in.}{2\ ft.} = \frac{3\ in.}{?\ ft.}$$

A proportion which represents scale drawing data

At this point we are dealing with a proportion problem exactly like the ones we worked with in the last section. The only purpose the drawing will serve for us is a way of verifying that our answer at least makes some sense.

Recall from the previous section that we can solve this particular problem in two ways. We can note that the denominator of the left fraction is twice its numerator, so the same will have to be true of the fraction on the right. The denominator must be 6, with a unit of feet which is how we set up our proportion. We could also note that if the numerator on the right is three times the numerator on the left, the denominators will have to follow the same pattern. Again, we get the same answer of 6.

Recall that we can compare the cross products of the proportion to confirm that it is valid. The cross product across each diagonal is 6, which proves that we have a valid proportion. If the cross products were not equal, we would know that we made some mistake.

WORKING WITH "NON-OBVIOUS" PROPORTIONS

Many proportion problems are not as simple as the ones that we've been working with. The numbers may not be such that the problem can be solved using basic computation, and the missing piece of information may end up being a decimal number. Such problems must be solved using basic algebra techniques which you will learn about in later math, and in the next book of this series. The topic of proportions is certainly revisited in the study of algebra, but for now just be sure to understand the general concept. Many exam questions are based on nothing more than that.

SO NOW WHAT?

Before progressing to the next chapter, it is absolutely essential that you fully understand all of the concepts in this one and in the previous chapters on fractions. In later math such as algebra you will do much more work with fractions, but the problems will be abstract in nature. This means that you must fully understand all of the concepts at this point while we are still working with simple numbers.

The next chapter introduces the concept of decimals which are very much related to fractions. If you don't fully understand fractions, the next chapter will probably be confusing and difficult for you.

CHAPTER NINE

Working with Decimals

WHAT IS A DECIMAL NUMBER?

A **decimal number** is a number that has a fractional component. Stated another way, it means that the number does not just represent a given quantity of wholes, but also represents some parts of a whole. For example, we could think of the number 3 as representing three whole pizza pies, regardless of how many slices each one is cut into. We could think of the number 2.5 as representing two whole pizza pies (again, with no regard to how they are cut), and also one-half of an additional pizza pie. You probably know from everyday experience that 0.5 equals ½, and in this chapter the mathematical reasons for that will be made clear.

EXTENDING THE PLACE VALUE CHART TO DECIMAL PLACES

In Chapter Three we worked with a place value chart for whole numbers. In this chapter we'll extend the place value chart to the right to accommodate places for decimal numbers. We use a **decimal point** (a period) to separate the whole number and the decimal place values. In a general sense, the decimal place values are the "mirror image" of the whole number place values. However, some important explanation is required.

Recall that as we move to the left in the whole number place value chart, each place has a value that is 10 times the value of the place on its right. For example, we have ones, tens, hundreds, thousands, and so forth. Conversely, as we move to the right, each place is one-tenth the value of the place on its left. This pattern continues on the right side of a the decimal point.

The concept of decimal place value is easiest to think about in the context of money which we deal with each day. We say $2.40 as "two dollars and forty cents," but we know that the 40 cents could made up of 4 dimes. We also know that each dime is worth 10 cents, and that 10 cents is one-tenth of a dollar since it takes 10 dimes to equal a dollar.

If we ignore the dollar sign and the 0, we are left with the decimal number 2.4 to examine. In a non-money context, many people pronounce that number as "two-point-four" which is fine. However, it is important to understand that in math, that value represents two wholes, and four tenths of another whole. You could think of it as two whole pizza pies, and 4 slices from another pie which has been cut into 10 equal slices (tenths).

We know that we group our dimes to the right of the decimal point, and that dimes represent tenths of a dollar. The place value immediately to the right of a decimal point is called the tenths place.

We know that in the amount $6.78, the 8 represents 8 pennies (cents). Each penny is one-hundredth of a dollar since it takes 100 of them to make one dollar. It also takes 10 pennies to make one dime, which means a penny is one-tenth of a dime. Converting our money example to the decimal number 6.78, we can see that the number represents 6 wholes (units), 7 tenths, and 8 hundredths.

Study the following place value chart to see how the pattern continues. No matter where we start on the chart, as we move left we multiply the current place value to 10, and as we move right we divide by 10. Note that there is no "oneths" place, and that the places to the right of the decimal point all end with "ths."

Thousands	Hundreds	Tens	Ones / Units	DECIMAL POINT	Ten*ths*	Hundred*ths*	Thousand*ths*
2	3	4	5	•	6	7	8
2,000	300	40	5		0.6	0.07	0.008

Place value breakdown of the number 2,345.678
(Two thousand, three hundred forty-five *and*
six hundred seventy-eight hundred*ths*)

WHY DO DECIMAL PLACE NAMES END WITH "ths"?

The simplest answer to this question is that there must be some way to distinguish if a column in the place value chart refers to a whole number, or to a fractional (decimal) value. Any place that ends with "ths" refers to a fractional value. For example, the number 3 in the hundreds place is worth 300. The number 3 in the hundred*ths* place (represented as 0.03) is worth 3/100.

WHERE IS THE "ONEths" PLACE?

When students see that the right side of the place value chart is, in a sense, the mirror image of the left side, a common question is, "Where is the *oneths* place?" The simple answer is that there is no such place, but it is important to understand why.

As we move to the left on the place value chart, each column is worth 10 times the value of the column on the right. As we move to the right, each column is worth 1/10 as much as the column on the left. If we think of the ones place as dollar bills, what would be 1/10 as big? The answer is a dime, or ten cents, or one-tenth of a dollar. That is why the tenths place is directly to the right of the ones place, of course with the decimal point in between since we've crossed the line between whole and fraction.

ADDING AND SUBTRACTING DECIMAL NUMBERS

As explained in the Introduction, this book does not go into details about doing arithmetic calculations by hand, except when doing so is straightforward, and serves to review important math concepts. As such, it is worth taking a quick look at how to add and subtract decimal numbers by hand.

The simple rule to remember is to line up the decimal points, and line up the place values on top of each other. This is just like what we did when we added whole numbers such as 123 + 45. The 4 had to go directly under the 2, since each of those numbers are in the tens place.

Let's add the decimal numbers 12.3 + 45.67 as shown in the example at left. Notice how the decimal points are arranged directly on top of each other. It is important to see that the 3 in the first number has not been "pushed over" to the right, since it represents 3 tenths. Instead, an "imaginary" 0 will serve in the hundredths place of the first number so that we'll have something to add to the 7 hundredths in the second number. That concept is discussed later in this chapter.

$$\begin{array}{r} 12.30 \\ +\,45.67 \\ \hline 57.97 \end{array}$$

Once the problem has been set up properly, the values can be added just like we did with whole numbers. When necessary, values may be carried to the place on the left. An example of this in the context of money would be converting 13 pennies into 1 dime and 3 pennies. Recall that our imaginary cash register was only able to hold nine bills in each compartment. We could now extend that concept and say that each change compartment can only hold nine coins of its type, and once a tenth coin is obtained, the coins needs to be exchanged for the next higher denomination value.

Subtraction of decimal numbers follows the exact same rules. Just remember to line up the decimal points and all of the place values. Placeholder zeroes can always be added on the right as necessary which is elaborated on next. Note that when adding and subtracting decimal numbers, the decimal point and place values in the answer will line up with all of the involved numbers.

WHAT IS THE DIFFERENCE BETWEEN 2.3 AND 2.03?

Zeroes in decimal numbers can be a bit confusing. The question to ask yourself is whether a particular zero is serving as a necessary placeholder since there are other digits to the right of it. For example, there is a big difference between 2.3 and 2.03. With the first number we are representing two wholes and three tenths. A money example would be two dollars and three dimes. With the second number we are representing two wholes and three hundredths. A money example would be two dollars and three cents. The zero in the second number is not optional. We need it to serve as a placeholder which in turn serves to "push" the three hundredths over into their proper column. We can't just leave a blank space where the 0 is. We must make it clear that the second number has no tenths, but it does have hundredths.

WHAT IS THE DIFFERENCE BETWEEN 2.3 AND 2.30?

The next logical question is what the difference is between 2.3 and 2.30. The answer is that there is no difference. Both numbers represent two wholes and three tenths. There are no **significant digits** to the right of the three. It doesn't matter that the second number doesn't have any hundredths. That doesn't change the value of the actual number. It's as though all you have in your pocket is two dollars and three dimes. The fact that you don't have any pennies doesn't change how much money you have, so there is no reason to mention it.

We say that the number 2.3 is a **terminating decimal**. It effectively stops at the 3. We can include zeroes to show that we also have no hundredths, no thousandths, and so on, but it doesn't change the number's value at all so we typically just omit them. When referring to money, though, it is customary to put a 0 in the hundredths (cents) place even if there are none.

WHAT IS THE DIFFERENCE BETWEEN 0.5 AND .5?

This is another common question among math students. The answer is that there is no difference. Both of those numbers have no whole number part, and have three tenths. The reason why the first form is preferable is simply because it calls more attention to the decimal point. It makes it very clear that there is no whole number component. With the second form, it is easy to wonder if the decimal point is simply a spot of stray ink on the page, if you even notice it at all.

WHY ISN'T THERE A DECIMAL POINT AFTER AN INTEGER?

This is yet another common question. It is important to understand that 17 and 17.0 are equal. However, unless we have a special reason, we usually just omit the decimal point and any decimal place zeroes when a value doesn't have a decimal component. With that said, it is still a good idea to have the sense that there really is an "invisible" decimal point to the right of every whole number. That comes into play for some concepts as we will see later in this chapter.

USING "AND" WHEN WRITING OR SAYING DECIMAL NUMBERS

This is a minor point, but you should be aware that when we write or say a decimal number in a math context, we use the word "and" to represent the decimal point. For example, we say or write 7.9 as "seven and nine-tenths." Admittedly, in everyday life it much more common to say "seven point nine."

There is no other circumstance in which we say "and" as part of a number. This means that if you see or hear the word "and," you know that everything that follows is the decimal portion of the number. For example, it is incorrect to write or pronounce 103 as "one hundred and three," even though most people do. That number is pronounced "one hundred three," with no "and." Compare that to the number 100.3 which is written and pronounced as "one hundred and three-tenths."

MORE ABOUT WRITING AND SAYING DECIMAL NUMBERS

It is important to be able to properly say decimal numbers and recognize them in print. We already learned that we use the word "and" to represent the decimal point. We then look to see how far to the right the number "reaches." We will use the smallest-valued decimal place as the word that we say or write.

For example, the number 7.32 extends into the hundredths place. We say "7" to represent the whole number, and "and" to represent the decimal point. What's confusing is that when writing or saying a decimal number, we treat the remaining digits as though they comprise a whole number, but we follow them up with the place value that we determined. In this case we say or write "seven and thirty-two hundredths." For the number 123.456 we say or write "one hundred twenty-three and four hundred fifty-six thousandths."

For decimal numbers that do not have a whole number component such as 0.37, we typically do not say "zero," nor do we say the word "and." The number 0.37 would be said or written as simply "thirty-seven hundredths." Just make sure that you understand the general idea of this concept so that you aren't tripped up if you are faced with a word problem in which the values have been written with words instead of numerals. Also make sure you understand that decimal point digits are not to be interpreted as though they represent a whole number, except when we are writing or pronouncing them. Admittedly this is confusing , but should be less so after reading the next section which discusses how we compare decimal numbers to determine which is greater.

COMPARING DECIMAL NUMBERS

The place value chart makes it easy for us to compare decimal numbers to see which of two is greater. We start our comparison in the tenths place, which is the most significant decimal place value. If that is not sufficient to determine which number is greater, we move to the right to the hundredths place, then continuing even further to the right if necessary. This is best explained by example.

The number 0.7 is greater than 0.429. We start our comparison in the tenths place. The first number has a 7 in that place, and the second number has a 4 there. That is all that matters. That proves that the first number is greater than the second. We are done. There is no need to examine any other places. Don't say or think, "But 429 is larger than 7," because that is only true when those numbers are functioning as whole numbers. In this example those numbers are serving as decimal place digits, which is a totally different context.

As another example, 0.638 is less than 0.65. We start our comparison in the tenths place. In both numbers that place has a 6, so we are in a "tie" situation. We must move to the right to the hundredths place to try to "break the tie." The first number has a 3 in that place, and the second number has a 5 there. That is all that matters. That proves that the first number is less than the second. The tie has been broken. We are done. There is no need to examine any other places. Don't say or think, "But 638 is larger than 65," because that is only true when those numbers are functioning as whole numbers. In this example those numbers are serving as decimal place digits, which is a totally different context.

When comparing decimal numbers, be sure to remain in the mindset of starting the comparison in the tenths place, and moving to the right only if necessary to "break ties." Do not at any time treat decimal place digits as though they are whole numbers.

Note that the whole-number portions of decimal numbers always take precedence over the decimal portions. What that means is that the number 4.2 is greater than 3.9999. Four wholes are more than three wholes. That is all that matters. It proves that the first number is greater than the second. We are done. There is no need to examine the decimal places. Don't say or think, "But

9999 is larger than 2," because that is only true when those numbers are functioning as whole numbers.

In the last example, it doesn't matter how many decimal places are to the right of the 3 in the second number, or what digits are in those places. There is nothing that can be done to make the 3 larger than 4 or even equal to 4. The decimal portion of 0.9999 simply increases the 3 so that it is slightly less than 4, but it is still less.

CONVERTING DECIMAL NUMBERS TO FRACTIONS

Our place value system makes it easy to convert decimal numbers into equivalent fractions. All we do is look to see how far to the right the decimal digits reach, and we use that place to represent the denominator of our fraction. That is similar to what we did when determining how to write or say a decimal number. The decimal digits become the numerator of our fraction, and we say the resulting fraction in the same way that we would say the decimal number from which it was derived.

For example, to convert the decimal number 0.73 to an equivalent fraction, we take note of the fact that the decimal digits extend into the hundredths place. That means that the denominator of our fraction will be 100.

The numerator will be 73, which is just the decimal digits. The resulting fraction is 73/100 which we say as "seventy-three hundredths," just like we say 0.73.

As another example, the decimal number 0.907 converts to an equivalent fraction of 907/1000. Again, we take note of the fact that the decimal digits extend into the thousandths place. That means that the denominator of our fraction will be 1000. The numerator will be 907, which is just the decimal digits. We say the resulting fraction as "nine hundred seven thousandths," just like we say 0.907.

CONVERTING FRACTIONS TO DECIMAL NUMBERS

Converting a fraction to a decimal number is very easy as long as you remember that a fraction is really a division problem—top divided by bottom. You also must remember that the order is significant. The fraction a/b is really $a \div b$, as we learned in Chapter Six.

$$a/b = a \div b = \frac{a}{b} = b\overline{)a}$$

Four equivalent ways of writing "*a* divided by *b*"

What this means is that to convert a fraction to a decimal, all we need to do is compute what the fraction literally means—top divided by bottom. As mentioned, this book assumes that you typically do most of your computations on a calculator. You must be careful to enter the numbers in the proper order. The top number must be keyed in before the bottom number.

For example, let's convert 7/10 to a decimal number by computing 7 ÷ 10. We get 0.7. Think about why that makes sense. We just learned that 0.7 is "seven tenths," which is exactly what our fraction was. Note that if we wrongly performed the computation in reverse as 10 ÷ 7, we would get an incorrect answer of approximately 1.43.

Let's convert 13/25 to a decimal number. We compute 13 ÷ 25 for an answer of 0.52. Let's see if that makes sense. We know 0.5 is equal to five-tenths which reduces to ½ (one-half). This means that 0.52 could be thought of as slightly larger than one-half. In the original fraction, the numerator of 13 is slightly larger than half of the denominator of 25, which means that the decimal equivalent which we computed makes sense. If we incorrectly did the computation in reverse as 25 ÷ 13, we would get a wrong answer of approximately 1.92.

While there are other methods for converting fractions to decimal numbers, one of which is discussed next, using this method will always work, and is especially easy if you are permitted to use a calculator. Just remember that a fraction is simply a division problem—numerator divided by denominator.

Note that there are many times when such a computation will result in a repeating decimal digit that "goes on forever" such as 0.33333..., or digits that repeat themselves in a pattern such as "0.789789789..." These situations are discussed in the next chapter.

ANOTHER METHOD FOR CONVERTING FRACTIONS TO DECIMALS

In order to review several of the concepts prevented, it is worth exploring another method of converting fractions to decimals which can be done in some cases.
Recall that we are always free to multiply the top or bottom of a fraction by a chosen number, as long we do the same thing to the other part of the fraction. Think about the fraction 7/20. Certainly we can convert this fraction to a decimal number by doing the procedure we just learned, which is computing top divided by bottom.

However, we could also note that if we were to multiply the denominator of the fraction times 5, it would become 100. We would of course have to do the same thing to the numerator, and we'd end up with the equivalent fraction of 35/100. The reason why this helps us is because the denominator is now a **power of 10** (e.g., 10, 100, 1000, etc.). Because of our place value system, it is now easy to convert the fraction to a decimal number. We know that 35/100 is "thirty-five hundredths." To write that as a decimal number, all we have to do is write 35 so that it ends in the hundredths place as 0.35.

As another example, think about the fraction 19/250. We could note that if we multiply top and bottom by 4, we'll end up with the fraction 76/1000. To write that fraction as a decimal, all we must do is write 76 so that it ends in the thousandths place as 0.076. Again, though, we could also just go straight to the the original procedure of computing top divided by bottom, or 19 ÷ 250, and we would also end up with the same answer of 0.076.

SO NOW WHAT?

Before progressing to the next chapter, it is absolutely essential that you fully understand all of the concepts in this one. The next chapter introduces some more advanced decimal topics that are very important. If you don't fully understand this chapter, the next chapter will probably be confusing and difficult for you.

CHAPTER NINE AND FIVE-TENTHS

More Topics in Decimals

REPEATING DECIMAL NUMBERS

As mentioned, when we convert a fraction to a decimal by computing top divided by bottom, we sometimes end up with what is known as a **repeating decimal**. A simple and common example is the decimal equivalent of the fraction ⅓. If you compute 1 ÷ 3, either on your calculator or by hand, you will see that the decimal digit 3 repeats itself "forever" as 0.3333... Of course your calculator will stop displaying digits when it runs out of room, and for some numbers your calculator may round the final digit of its display, but it is still the case that we are dealing with a repeating decimal.

To represent a repeating decimal digit, we write a bar over the digit that repeats. For example, the decimal equivalent of 1/3 would be written as $0.\overline{3}$. That is not at all the same as 0.3 without the bar which is the decimal equivalent of 3/10, which in turn is not equal to 1/3.

Some decimal numbers are characterized by the repetition of several digits which form a pattern. For example, if you convert the fraction 357/999 to a decimal number by computing top divided by bottom, you will get 0.357357357... The pattern continues even if your calculator runs out of room and/or adjusts the final digit due to rounding. We would represent the above decimal number with our bar notation as $0.\overline{357}$.

Just be sure to not get flustered by the concept of repeating decimals, or by the bar notation. You will work with repeating decimals more in later math.

TERMINATING DECIMALS

Many decimal numbers simply stop, meaning if we were to keep computing more digits using division, all we would get is more and more zeroes. We've already seen that 0.7 is the same as 0.70, 0.700, 0.7000, and so on, so we could say that the decimal number terminates at the 7. We say that 0.7 is a **terminating decimal**. It doesn't have any significant digits that repeat themselves, nor do the decimal digits continue endlessly without repeating in some type of pattern.

NON-REPEATING DECIMALS

A third, special type of decimal is known as a **non-repeating decimal**. Some decimal numbers do not terminate, but yet they do not have any digits that repeat in a pattern like the examples we've seen.

The most common example of a non-repeating decimal which you will work with extensively in later math is the number pi (π). The value "three-point-one-four" may come to mind based on your previous experience with π, but the reality is that its decimal digits go on and on forever without repetition or pattern. For now, just make sure you understand the concept that a decimal number will either be terminating, repeating, or non-repeating.

COMPARING FRACTIONS USING REFERENCE FRACTIONS

This topic may seem out of place in a chapter on decimals, but the reasons for such will be explained shortly. Sometimes we are asked to compare fractions to determine which of two is bigger, and sometimes we are asked to sort a list of fractions from smallest to largest.

This is often easily done by making use of what are called reference fractions. A **reference fraction** is simply a fraction whose value you know and are familiar with. They serve as "guideposts" relative to other fractions.

For example, you probably know that ¼ is smaller than ½. Think about those fractions in regard to a pizza pie. Now think about the fractions 7/51 and 65/82. We don't know the precise value of those fractions, although we could certainly convert each one to a decimal as we learned. However, such a step is not necessary.

It is easy to see that the fraction 7/51 is certainly smaller than ¼. 7/28 is ¼, but the denominator of 51 is larger which we learned makes the fraction smaller. We can use the same logic to deduce that 65/82 is certainly bigger than ½. That is all we need to know. From this is it easy to see that the first fraction is smaller than the second one, although we don't know exactly by how much.

Make sure you understand the general concept. Next we'll review how to convert fractions into decimals in situations where two fractions seem so close in value that they can't be distinguished by way of reference fractions. In the next chapter we'll look at a chart of common reference fractions and their decimal equivalents.

ARRANGING "NON-OBVIOUS" FRACTIONS IN ORDER FROM LEAST TO GREATEST

The reason why this topic is in the chapter on decimals and not fractions is because it is often the case that we need to turn to the decimal equivalents of fractions in order to accurately compare them.

Again, remember that it is easy to convert any fraction into a decimal by computing top divided by bottom. Also remember that is it easy to compare decimal values to one another by starting the comparison in the tenths place, and moving to the right as needed in order to "break ties." This technique is important in cases where it is not practical to make use of reference fractions like we just learned.

An example would be if wanted to compare 108/429 to 517/2061. Both of those fractions appear "to the eye" to be roughly equivalent to ¼. However, both would need to be converted to decimals in order to determine with accuracy which has the smaller and the larger value. In doing so, we would see that the first fraction is equal to about 0.252, and the second fraction is equal to about 0.251, proving that the first fraction is slightly larger.

SHORTCUT FOR MULTIPLYING BY POWERS OF 10

Recall that examples of powers of 10 are 10, 100, 1000, etc. These values are obtained by raising a base of 10 to various powers (exponents) as we previously learned.

Powers of 10 are very common in math because our entire place value system is based on them. As such, it is important to memorize some very basic shortcuts which make it easy for us to quickly multiply by powers of 10.

To multiply a number times 10, simply move the decimal point one place to the right, remembering that whole numbers have an "invisible" decimal point on the right. For example, 7.46×10 is 74.6. We moved the decimal once place to the right. As another example, 45×10 is 450. We had to add a zero in order to create a space for the "invisible" decimal point to move into. Then, since that "invisible" decimal point ended up at the right of the number, we made it "invisible" again.

To multiply a number times 100, just move the decimal point two places to the right. Again, add zeroes if you run out of digits but you still need to move the decimal point more places. For example 8.7×100 is 870. The first move to the right got us to 87. We then needed to add a

0 in order to move the decimal point a second time. We ended up with 870, followed of course by an "invisible" decimal point that we don't need to write.

To multiply a number times 1000, just move the decimal point three places to the right, following all of the above procedures. As expected, the pattern continues. To multiply a number times 10,000 you would move the decimal point four places to the right. This idea will be important in the upcoming section on scientific notation.

SHORTCUT FOR DIVIDING BY POWERS OF 10

As you may predict, the shortcut for dividing by powers of 10 works the same way as multiplying by powers of 10 except we move the decimal point left instead of right. This makes sense because when we multiply we want to end up with a larger number, and when we divide we want to end up with a smaller number.

To divide a number by 10, just move the decimal point one place to the left. To compute 17 ÷ 10, just move the "invisible" decimal one place left to get 1.7. To divide 0.8 by 10, move the decimal to the left one place to get 0.08. Recall that the 0 in the units place is not necessary, but is usually added for clarity.

To divide a number by 100, move the decimal point two places to the left, adding zeroes as needed. For example, to compute 0.3 ÷ 100, move the decimal left once to get 0.03, and then again to get 0.003. We had to add an extra zero so that the decimal would have a place to move to.

As expected, the pattern continues for higher powers of 10 in the same way that it does for multiplication. It is important to feel comfortable with these shortcuts since you will be able to make use of them very frequently. Do some experimenting on your calculator to convince yourself that they will always work as described.

DECIMALS WITHIN FRACTIONS

Recall that we informally described a fraction as simply a value divided by a value. Decimal numbers are just special types of values which means that there is nothing wrong with a decimal number comprising either the numerator or denominator or both parts of a fraction.

To convert such a fraction into an actual decimal number, you can still follow the procedure of computing top divided by bottom. If you are doing a problem which requires that the fraction not contain any decimal values, remember that it is always OK to multiply both top and bottom of the fraction by the same value of your choosing. As we've seen, it is easy to multiply a value times a power of 10 in order to "move" the decimal point over to the right such that it is effectively removed.

For example, if we wanted to "remove" the decimal points from the fraction 7.4/28.56, it would be necessary to multiply both top and bottom of the fraction by 100. This would move each decimal point two places to the right. We would end up with an equivalent fraction of 740/2856, which is comprised of only whole numbers.

The main point to understand is that a decimal number in a fraction should not be cause for concern. Just handle the fraction in the same way that you would any other.

ROUNDING DECIMAL NUMBERS

In Chapter Three we learned how to round whole numbers to various places. Be sure to review that section in that chapter if you don't fully remember the procedure. Rounding decimal numbers works in exactly the same way. We need to round decimal numbers more often than whole numbers since decimal numbers can often "go on forever" as we've seen. It is usually practical to round them to a given place in an effort to artificially terminate them. Certainly when we are dealing with money we almost always round dollar amounts to the nearest penny (hundredth).

Let's review the rounding procedure that we learned in Chapter Three, and apply it to a decimal number. Let's round 127.483 to the nearest hundredth. Be very careful in noting that the problem references the hundred*ths* place, and not the hundreds place. We'll call the hundredths place our "target place," and bold it. To round to a given place, we always examine the place to the right of the target place. We'll call the digit in the place on the right the "check digit," and underline it. In the example we have 8 as the target digit, and 3 as the check digit. Let's review the first rule of rounding:

If the check digit is 4 or lower, the digit in the target place *remains as is*, the check digit becomes a 0, and all the digits to the right of it become 0. In this example, 127.483 rounded to the nearest hundredth is is 127.480, which in turn becomes 127.48 because of what we learned about zeroes at the end of decimal numbers.

$$127.4\mathbf{8}\underline{3} \rightarrow 127.480 \rightarrow 127.48$$

Rounding 127.483 to the nearest hundredth

Now let's round that same number to the nearest tenth. We now have 4 in the target place, and 8 as the check digit. Recall the second rule of rounding:

If the check digit is 5 or higher, the digit in the target place is increased by 1, and the check digit and all the digits to its right become 0. In this example 127.483 rounded to the nearest tenth is 127.500 which in turn becomes 127.5 because of what we learned about zeroes at the end of decimal numbers.

$$127.\mathbf{4}\underline{8}3 \rightarrow 127.500 \rightarrow 127.5$$

Rounding 127.483 to the nearest tenth

Let's round 239.514 to the nearest integer (whole number). Our target place is the ones place which contains a 9. The check digit which is in the tenths place is 5. Since the check digit is 5 or higher, we must increase the digit in the target place by 1, and make the check digit and everything to the right of it 0. But in this case, how do we increase the target place digit of 9? Since we can't make it 10, we make it 0 and effectively carry the 1 into the tens place so that the 3 is increased to 4. In this example, 239.514 rounded to the nearest integer is 240 after dropping all of the decimal digits which became zeroes, and the decimal point itself which is no longer necessary.

239.<u>5</u>14 → 240.000 → 240

Rounding 239.514 to the nearest integer

Be careful because sometimes we are asked to round a decimal number to one of the whole number places, even though it happens to be a decimal number. An example would be rounding the number 1234.567 to the nearest hundred, but not hundred*th*. Just follow all of the rules that you learned for rounding.

SCIENTIFIC NOTATION

Scientific notation is a special format that is used to represent numbers that are either extremely large or extremely small. This is because such numbers typically contain a large quantity of zeroes either to the left or right of the decimal point, and it is impractical to have to count how many there are.

In scientific notation, numbers are written in the form $a \times 10^b$. The "*a*" portion must be a value that is greater than or equal to 1, but less than 10. It contains the significant (non-zero) digits of the very large or very small number that we are representing. Some valid values of *a* are 7.2, 6.385, 1, and 9.9999. Some *invalid* values of *a* are

10, 47, and 0.123 since each of those values is either less than 1, or is greater than or equal to 10.

$$a \times 10^b$$

The format for scientific notation

The value of *a* is multiplied by the 10 (it's always 10) raised to a given power, which we'll denote "*b*." The exponent "*b*" can either be positive or negative. We haven't yet learned about negative exponents, but there is only one simple rule that you must learn for this topic.

All you need to know is that the exponent *b* tells us how many places we must move the decimal point in the value *a*. If *b* is positive, we move the decimal the given number of spaces to the right, and if it's negative we move the decimal point the given number of spaces to the left. Recall that we learned about how to do this in the section of this chapter that covered the shortcut for multiplying and dividing by powers of 10.

Here is an example. Let's convert 437,000,000 to scientific notation. We must take the significant digits 437 and turn them into a value that is greater than or equal to 1, but less than 10. We can do that by inserting a decimal point within those digits to get 4.37. That will be the "*a*" value in the scientific notation version of the number.

We now must determine how many times we would need to move the decimal point to the right in order to get back our original number. If you count carefully, you'll see that it is 8. Remember that we add zeroes as needed so that the decimal point will have places into which to move. This means that the "*b*" value in the scientific notation version of the number will be 8, and it will be positive since we need for the decimal point to move to the right. This means that 437,000,000 can be written in scientific notation as 4.37×10^8. If we started with the scientific notation and we needed to convert it to traditional format, we would just reverse the steps.

Now let's try representing a very small number such as 0.0000007302 in scientific notation. The significant digits are 7302. We do need to include the 0 which is between the 3 and the 2 since it's serving to hold a place between two significant digits, but we don't include any of the other zeroes. It is certainly true that they are also holding places, but they will not serve any purpose in determining our "*a*" value which is the next step. As before, we need to insert a decimal point within the significant digits to get a value that is greater than or equal to 1, but less than 10. Our only option is 7.302.

We now must determine how many times we would need to move the decimal point to the left in order to get back our original number. If you count carefully, you'll see that it is 7. Remember that we add zeroes as needed so that the decimal point will have places into which to move. This means that the *b* value in the scientific notation version of the number will be -7. It is negative since we need for the decimal point to move to the left. This means that 0.0000007302 can be written in scientific notation as 7.302×10^{-7}. If we started with the scientific notation and we needed to convert it to traditional format, we would just reverse the steps.

SO NOW WHAT?

Before progressing to the next chapter, it is essential that you fully understand all of the concepts in this one. The next chapter introduces percents which are simply yet another form that fractions and decimals can take. If you don't fully understand this chapter, the next chapter will probably be very confusing and difficult for you.

CHAPTER TEN

Working with Percents

WHAT IS A PERCENT?

Like fractions and decimals, a **percent** is yet another way of representing part of a whole. Depending on what we are working with or talking about, one of those three options might make more sense than the others.

As an example, later in this chapter we will learn that 30% is equal to 0.3, which in turn is equal to 3/10. Those are three different yet equivalent ways of representing the same part of the same whole. However, in the context of a store sale, the percent version is what we would choose, and what people would expect to see.

In a general sense, a percent is simply a benchmark. If a store is having a "30% Off" sale, it means that you will save \$30 on a \$100 item. Because our math system is based on groups of 10 as we've seen, 100 is a convenient number to work with. We can easily deduce that a \$50 item at the store will be reduced by \$15—half that of the \$100 item. We may not know exactly how much money we will save on a \$69.95 item, but the benchmark of saving \$30 on a \$100 item gives us a general idea of the discount which the store is offering. Later in the chapter we'll learn how to do precise calculations with percents.

Percents are based on the place value of hundred*ths*. We use the % symbol to represent a percent. It is very important to understand that the % symbol simply translates into "over 100." For example, 57% simply means "57 over 100," or 57/100, or fifty-seven hundredths. This is represented symbolically below.

$$n\% = \frac{n}{100}$$

The symbolic representation of a percent

If you see a percent sign after a number, and you need to do any computations with it, first drop the % sign, and then put the number over 100 regardless of what the number is. You will then likely convert it to a decimal, discussed later in the chapter.

CONVERTING FROM A PERCENT TO A FRACTION

Although this topic has just been discussed, some special points must be made. The word "percent" literally means "out of 100." The % symbol can be thought of as "over 100." It is a shorthand notation used when we want to save space, or express a portion of a whole in a more "everyday" manner. As mentioned, if a store is having a "30% Off" sale, they will not advertise it as "30/100 Off" even though that means the same thing.

To convert from a percent to a fraction, we first put the given value over 100 regardless of what it is, and then we will later simplify the fraction if necessary. The value in question could be a simple integer like 25, or it could be a decimal number less than 1 such as 0.5, or it could be a value greater than 100 which may seem strange. Whatever it is, just put it over a denominator of 100 and remove the % sign. Some examples are shown below:

$4\% = \frac{4}{100}$	$25\% = \frac{25}{100}$	$0.07\% = \frac{0.07}{100}$
$150\% = \frac{150}{100}$	$300\% = \frac{300}{100}$	$8.4\% = \frac{8.4}{100}$

The % symbol just means "over 100"

What is done next depends on the calculation being performed, or the context of the particular problem. In many cases it will be best to reduce the fraction to lowest terms. Remember that when converting from a percent to a fraction, don't perform any intermediate steps. Just take the given value and put it directly over 100, then drop the % sign.

CONVERTING FROM A PERCENT TO A DECIMAL

It is easy to convert from a percent to a decimal. We typically do this when we need to do some type of computation involving the percent, since a decimal is better suited for such.

We've seen that to convert from a percent to a fraction, just put the given number over 100 and drop the % sign. To convert that fraction to a decimal, we must remember that a fraction is actually a division problem. "Over 100" means to divide by 100. We've seen the shortcut where we can quickly divide by 100 by moving the decimal point two spaces to the left. Remember that all whole numbers really have an "invisible" decimal point on the right. See the next example.

$$47\% = \frac{47}{100} = 0.47$$

Converting a percent to a fraction and then to a decimal

Of course the step in the middle could really be skipped. Since you know that the % symbol means to effectively divide by 100, you can just jump straight to doing that if you need the percent to be converted to a decimal.

It is important to understand that this same procedure should be followed regardless of the value in front of the % sign. Some examples are below:

$83\% = \frac{83}{100} = 0.83$	$0.9\% = \frac{0.9}{100} = 0.009$
$120\% = \frac{120}{100} = 1.2$	$300\% = \frac{300}{100} = 3$

Examples of converting percents to decimals via fractions

CONVERTING FROM A DECIMAL TO A PERCENT

We have just seen that to convert from a percent to a decimal, the procedure is effectively to drop the % sign and move the decimal two places to the left since the % sign really means to divide by 100.

To convert from a decimal to a percent, the steps can simply be reversed. First, move the decimal two places to the right (which is to effectively multiply by 100), then add the % sign. Some examples follow:

$0.375 = 37.5\%$	$0.001 = 0.1\%$
$4.5 = 450\%$	$2 = 200\%$

Examples of converting decimals to percents

CONVERTING FROM A FRACTION TO A PERCENT

It can be a bit trickier to convert from a fraction to a percent. Remember that "percent" means "over 100." That means that if you can convert the fraction to an equivalent one with a denominator of 100, you're all set.

As a simple example, it is very easy to convert 27/50 to have a denominator of 100. Just multiply both numerator and denominator by 2 which we've seen is allowed since 2/2 equals 1, and there is never any harm in multiplying a number by 1. We end up with 54/100, which we know is equivalent to 54%.

In most cases it will not be that simple. For example, in the fraction 41/95 there is nothing obvious that we can multiply top and bottom by to get a denominator of 100. Remember, it is multiplying that we have to do, not adding. We convert fractions to equivalent ones by multiplying top and bottom by the same number. Don't get confused and think that we can just add 5 to the numerator and denominator. That is not how it works.

To convert 41/95 to a percent, we would have to convert it to a decimal by computing top divided by bottom as we've learned. That gives us a decimal whose digits start with 0.43157... Recall that to convert a decimal to a percent, just move the decimal point two places to the right and add a % sign. As part of that step we would likely round to a given place value such as tenths or hundredths, ending up with 43.2% or 43.16% depending on which we chose.

IS IT A DECIMAL OR A PERCENT?

It is fairly common to see a value such as 0.5%. As expected, most students get flustered by this, and demand to know whether or not the given number is a decimal or a percent. Since the value includes a percent sign, we say that it is a percent. The particular percent in the example happens to involve a decimal number which is nothing to be concerned about. Remember that the % sign tells us to take the value that precedes it, place it over a denominator of 100, and then drop the % sign. That means that 0.5% is equivalent to 0.5/100.

It is important to understand the significance of a value like 0.5%. We know that 1% means 1/100, or 1 out of 100. Perhaps 1% of the population has a particular disease. Now think about the value 0.5 in compared to 1—it is half. That means that 0.5% can be thought of as "half of one-percent." Instead of 1 out of 100, it actually means 1 out of 200. We said that 0.5% converts to 0.5/100. We can multiply top and bottom by 10 to get an equivalent fraction of 5/1000, which reduces to 1/200.

$$0.5\% = \frac{0.5}{100} = \frac{0.5 \times \mathbf{10}}{100 \times \mathbf{10}} = \frac{5}{1000} = \frac{1}{200}$$

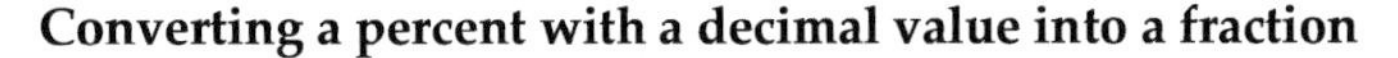

Converting a percent with a decimal value into a fraction

Recall from the previous chapter that we sometimes see decimal values within fractions, and they are nothing to be concerned about. Depending on what we need to do with the fraction, we may leave it as is, or we may choose to multiply both top and bottom by a power of 10 so that it becomes a fraction comprised of only integers.

PERCENTS THAT ARE GREATER THAN 100%

Just as we sometimes see percents that are less than 1%, we also sometimes see percents are that greater than 100%. Admittedly this can be a bit confusing. 100% means 100/100, which can be taken to mean one whole. Anything greater than that would imply a portion that is actually more than what you started with. While it may be common in everyday life to say something like "I gave it a 110% effort," there is really is no such thing, just as it is not possible to eat 110% of the cookies that are in a cookie jar. The most you can eat is all of them, or 100%.

The reason why we need percents that are greater than 100% in math is because they come up quite frequently when we do computations that involve percent increase. We'll learn about that later in the chapter, but as a quick preview, if the price of something doubles, we say that it increased by 100%. If the price of something increases by more than double, which can certainly happen especially over the course of many years, we say that the price has increased by a value that is greater than 100%.

For now, just remember that regardless of what value you see to the left of a percent sign, put that value over 100 and drop the percent sign.

EQUIVALENT PERCENTS / DECIMALS / FRACTIONS

There are some percent values that occur very frequently. It's best if you can memorize their decimal and reduced fractional equivalents so that you can save time and not have to "reinvent the wheel" each time you encounter them. See the chart below which you should try to memorize and be able to recreate on your own.

Percent	Dec.	Fract.	Percent	Dec.	Fract.
0%	0	0	25%	0.25	1/4
0.5%	0.005	1/200	33 ⅓%	$0.33\overline{3}$	1/3
1%	0.01	1/100	50%	0.5	1/2
2%	0.02	1/50	66 ⅔%	$0.66\overline{6}$	2/3
5%	0.05	1/20	75%	0.75	3/4
10%	0.1	1/10	100%	1	1
12 ½%	0.125	1/8	150%	1.5	3/2
20%	0.2	1/5	200%	2	2

Common percent / decimal / fractional equivalents

COMPUTING PERCENT OF INCREASE/DECREASE

A typical problem involves being given the original price of an item at some point in the past or present, along with the current or future price of the item which could either be higher or lower than the original price. The problem then asks us to compute the percent increase or decrease. This is easy to do as long as you remember this three-step procedure:

Step 1 of 3: Compute the difference between the original and the current prices. Just subtract the smaller of the two from the larger. That is the change in price.

Step 2 of 3: Divide the change in price by the original price. We always divide by the original price regardless of whether we are dealing with an increase or decrease.

Step 3 of 3: Convert your decimal answer to a percent by moving the decimal two places to the right and adding the % sign. Round according to any directions given.

$$\frac{\textit{change in price}}{\textit{original price}}$$

To compute percent increase or decrease, always take the change in price and divide it by the original price

In most cases the instructions will simply be to compute the percent change in price which means that you won't even have to make the distinction if it was an increase or a decrease. This is because it is obvious whether the price went up or down. In some cases we would use a negative percent to indicate a percent decrease, but the context of the problem and its instructions will make that clear. In these types of problems, all that is being tested is whether you understand the three-step procedure outlined on the previous page.

Let's try an example: Five years ago, the price of a postage stamp in Zafrania was 61 cents. Today, the price of a postage stamp in Zafrania is 94 cents. By what percent did the price increase in the last five years? Round the answer to the nearest whole number percent.

According to the procedure, we must first compute that the change in price was 33 cents. Next, divide the change in price by the original price of 61 cents. We compute $33 \div 61$ (not the other way around) to get about 0.541 (rounded). Convert to a percent to get 54.1% which we'll round to 54%—our answer. The price increased by a bit more than half of its original value.

Note that in this section we worked with matters of price which is common among related problems. However, the concept and procedures are exactly the same even if we are working with values that don't involve money.

PROBLEMS INVOLVING "PERCENT OF"

Many word problems involve computing a "percent of" or a "percent off" a number. These two things are not at all the same, although they look and sound very similar.

Let's first compute a percent *of* a number. A typical problem is, "What is 56% of 33?" Let's start by estimating the answer. You know that 56% is a bit more than 50%, and you know that 50% means ½. 33 is very close to 30, so let's work with that. Think about a class of 30 students of which half are boys. That would be 15. In this case we're dealing with a percent that's a bit more than half, and we're dealing with a number that's a bit higher than 30, so an estimate of 18 or 19 is reasonable.

Let's figure it out mathematically. Recall that whenever we see the word "of" between two values, it means to multiply. Our problem becomes 56% × 33. Also recall that to do computations involving percents, we should convert the percent to a decimal. Our problem is now 0.56 × 33. Multiply to get 18.48, which you would round according to any instructions given.

PROBLEMS INVOLVING "PERCENT OFF"

Now let's look at problems involving "percent off." These problems usually involve computing some sort of discount or price reduction. It is very important to read such problems carefully to determine if you are being asked to compute the amount of the discount, or the price after the discount has been deducted.

Here is a typical problem: "A $72 item is on sale for 25% *off*. How much money will you save? How much money would you actually pay for the item?" This problem involves both computing the discount as well as the final price. Many problems only request just one of those pieces of information.

The first step is to compute the amount of savings. A "25% off" sale means that 25% of the original price will be deducted from that original price. That means we must first compute 25% of the original price. In this case, we will use the "percent *of*" procedure described in the previous section. We compute 0.25 × $72 to get $18. That is how much money you will save on the item.

There are two ways to compute the actual amount that you will pay after the discount is applied. The first way is to simply subtract the amount of the discount from the original price. We compute $72 – $18 to get $54.

The second way is to understand that if the store will be taking 25% *off* the original price, you will be paying 75% *of* the original price. This is because the discount percent and the percent that you pay must always add up to 100%. The 75% came from computing 100% – 25%. We can compute 0.75 × $72 to get $54, which is the same answer that we got using the first method.

INCREASING OR DECREASING A VALUE BY A GIVEN PERCENT

Sometimes instead of being asked to compute the percent increase or decrease between two values like we did earlier in the chapter, we are instead given an amount and a percent, and are told to increase or decrease the given amount by the given percent.

As an example, a problem may state that a person's rent is $780 per month, and is soon going to increase by 3%. The problem will likely ask you to determine the person's monthly rent after the increase takes effect.

While there are some shortcuts for doing a problem like this, it is best to do it in two simple steps. The first step is to compute the amount of increase, and the second step is to add that increase to the original amount to get the new amount including the increase.

To find the amount of increase in this problem, we have to compute 3% of the current rent of $780. Recall that when we see the word "of" in between two values, it means to multiply. We'll convert 3% to its decimal equivalent of 0.03, and then multiply that times $780. We get $23.40 which is the amount of increase. Since the problem asked us to calculate what the monthly rent will be after the increase is applied, we simply have to add the amount of increase to the original amount, giving us a post-increase monthly rent of $803.40

Note that had the problem involved a decrease, we would subtract the amount of decrease instead of adding it, but we would compute the amount of decrease in the same way. Also note that sometimes a problem will ask us to only compute what the increase or decrease will be based on a given percent change. For such a problem it would be wrong to add or subtract the percent change to/from the original amount. Always read the problem carefully to determine what is being asked.

Also make sure you understand that in this section we learned how to handle problems in which we were given a value and a percent, and were told to increase or decrease the amount by that percent. That is different than the problems that we learned about earlier in the chapter in which we were given two values, and were asked to compute the percent of change between them.

PROBLEMS INVOLVING SALES TAX

Many problems ask us to compute the sales tax on an item. In some cases we are only asked to compute the amount of tax which will be paid, and in some cases we will be asked to compute the final price of an item after the sales tax has been applied. Always read the problem carefully to determine what is being asked.

Problems involving sales tax are solved in exactly the same way as any problem involving a percent increase such as the one we worked with in the previous section. Sales tax is simply a percent increase on an amount. Just follow the steps in the previous section.

For example, a problem may state that you want to buy a $51 item in a state whose sales tax rate is 8.25%. To compute the amount of tax you will pay, first convert the percent to a decimal to get 0.0825, and then multiply that decimal times the original amount of $51. We get $4.2075 which we round to the nearest penny as $4.21. If that was all the problem wanted, we're done. If the problem asked us to compute the final price of the item after the tax is applied, we would just add the $4.21 to $51 to get $55.21.

TWO OTHER COMMON MODELS OF WORD PROBLEMS INVOLVING PERCENTS

Many word problems involving percents follow the pattern of, "What is 31% of 119?" We learned how to do this. The word "of" in between two values means multiplication. We simply convert the percent to a decimal, translate "of" into multiplication, and then multiply the given values. We compute $0.31 \times \$119$ to get 36.89.

Let's estimate to see if the answer is reasonable. 31% is a bit less than 33% which we've seen is equal to about one-third (⅓). 119 is a bit less than 120. One-third of 120 is 40 (just divide it by 3). Since both the percent and the given value are a bit less than our test values, we should expect an answer that is a bit less than 40 which is what we got.

Another typical word problem involving percents is of the form, "46 is what percent of 59?" That problem could also be stated equivalently as, "What percent of 59 is

46?" Take careful note of the fact that unlike in the previous problems, we are not being given a particular percent. We are being asked to compute one. This problem is effectively asking us to do a comparison between 46 and 59. It wants us to determine what portion of 59 is represented by 46. Imagine a gathering of 59 people, of which 46 are women. What percent of the attendees were women? The question will probably include instructions for rounding your answer, perhaps to the nearest tenth or whole number percent.

To solve problems of this form, we must arrange the given numbers into a fraction. It's easy to then convert that fraction into a decimal and then into a percent, both of which we learned how to do. In this problem, we must arrange the given values into 46/59 to do our comparison. Remember that a fraction is really a division problem (top divided by bottom). We must compute 46 ÷ 59 to get 0.78 (rounded). Move the decimal two places to the right to convert that to 78% which is our answer.

"x is what percent of y?"

"What percent of y is x?"

$$\rightarrow \frac{x}{y} \text{ or } x \div y$$

Two equivalent formats of a common word problem pattern involving percents

SO NOW WHAT?

Before progressing to the next chapter, it is absolutely essential that you fully understand all of the concepts presented up to this point. They form the foundation of all the math that you will study from this point forward. If you don't fully understand everything that has been presented, your study of later math including algebra will probably be very confusing and difficult for you.

The next chapter introduces some basic concepts in probability and statistics. The good news is that most students find those topics to be somewhat fun and interesting. However, most standardized exams only include a few token questions on those topics, instead favoring the material presented in the other chapters.

CHAPTER ELEVEN

Basic Probability and Statistics

COMPUTING THE MEAN (AVERAGE)

When working with a list of values such as your exam scores for a semester, it is helpful to compute a single value which best represents all of the scores. We refer to that value as the **average**. Think of it as the balance point. It should be pulled up by high scores and pulled down by low scores. Scores that are extremely high or low should affect the average to reflect such. In math, we refer to a simple average calculation as the **mean**.

Computing an average (mean) is very simple. Just add up all of the scores involved, and then divide by the number of scores added. *Scores of 0 must be counted!*

Average = Sum of Scores ÷ Number of Scores

The formula for computing average (mean)

For example, someone's test scores for the semester were 94, 97, 0, 89, and 91. Note that there are 5 scores—the 0 counts. The sum of the scores is 371. Compute 371 ÷ 5 to get 74.2 which is the mean. If you were instructed to round your answer to the nearest integer, it would be 74. Notice how much the 0 lowered the average.

What if the teacher had a policy to drop the lowest grade? Then we would only be dealing with four scores. Add them to get the same 371, but this time divide by 4 to get an average of 92.75. What a difference!

COMPUTING THE MEDIAN

In a list of numbers, the **median** is defined as the middlemost value *after the list is arranged in ascending order.* For example, if you are given the list, 83, 45, 7, 29, 36, first re-sort the list as 7, 29, 36, 45, 83. The middlemost value is 36 which is the median. That is the procedure to follow if the list has an odd number of entries.

If the list has an even number of items, there will not be one specific number in the very center. For example, look at the already-sorted list 3, 5, 7, 8, 19, 28. For lists with an even number of items, we compute the median by taking the mean (average) of the two middlemost entries.

In the list above, the two middlemost entries are 7 and 8. Follow the instructions in the previous section to determine that 7.5 is the mean (halfway point) of those two numbers. We use that value as the median, even though it doesn't actually appear in the list.

MEAN VERSUS MEDIAN

Sometimes we're asked whether a given practical situation would better lend itself to a computation involving the mean or the median. When we compute the mean, a highly deviant score can significantly "pull" the mean higher or lower. For example, a high test grade can really pull up your average, and a low test grade can really pull it down. That is exactly how we want it to work—it's fair in both of those situations.

Now think about the situation in which a newspaper is reporting on the typical salary of residents in a small town. Imagine that virtually everyone in the town earns between $20,000 and $40,000 per year. Now imagine that some billionaire decides move to the town. Think about what that would do the mean income of the town's residents. It will greatly inflate the mean, and as such it would no longer be a reasonable representation of the economic status of the town's residents.

What we're really interested in is a number that paints an accurate picture. The statistic that we need for such a situation is the median, and that is why you will often read of the median income of a group rather than the mean or average.

In the last example, if we arrange the incomes of the town's residents in order from low to high, the middlemost value will probably be somewhere around $30,000. The fact that a single billionaire lives in the town does not in any way alter the median, but it does drastically alter the mean. Conversely, if a few people lose their jobs and are earning nothing, that doesn't unfairly pull down the median either, but it would pull down the mean since we'd be averaging in zeroes.

In situations involving test scores where we want each value to carry the same weight, we typically use the mean. In situations where we want an overall representation of central tendency, we typically use the median.

FINDING THE MODE

In a list of numbers, the **mode** is the number which occurs the most frequently. For example, in the list 24, 7, 13, 24, 85, the mode is 24. If a list has two or more sets of entries that are tied for the most frequently occurring, there will be more than one mode. For example, in the list 5, 5, 6, 6, 7, 7, 8, there are three modes: 5, 6, and 7. If all of the entries in a list occur the same number of times, then there is no mode. That would be the case for the list 1, 1, 3, 3, 5, 5, 7, 7.

FINDING THE RANGE OF A LIST OF NUMBERS

The **range** of a list of numbers is defined as the difference between the largest and the smallest entries. For example, in the list 17, 2, 36, 9, the range would be 34, obtained by computing 36 – 2.

BASIC CONCEPTS IN PROBABILITY

Probability is the study of how likely it is that an **event** will result in a particular **outcome**. We define "event" as whatever it is that we are doing or analyzing. Perhaps we are predicting the weather for tomorrow, or flipping a coin, or rolling a die, or one of those followed by the other. We define "outcome" as the result of an event.

We can express the chances of a given outcome in terms of a decimal, fraction, or percent depending on what is more convenient or appropriate. For example, a weather forecast might predict a 70% chance of rain. Recall that we could also express this as 0.7 or 7/10, but 70% makes the most sense for a weather report.

If it is impossible for an outcome to occur, we say that its probability is 0%. An example of an impossible event would be rolling a 7 on a single standard die. If an event is guaranteed to occur, we say that it's probability is 100%. An example of a guaranteed event is drawing a red marble from a bag that contains only red marbles.

Remember that 0% can be represented in fraction form as 0/100, which as a decimal would just simply be the number 0. 100% can be represented in fraction form as 100/100, which as a decimal is just the number 1. If something is equally likely to occur as it is to not occur, for example, a coin landing on heads, we say that its probability is 50%, or 0.5, or ½ (reduced from 50/100).

THE GENERAL PROBABILITY FORMULA

We define the **probability** of an event as the number of favorable outcomes divided by the number of total outcomes. The end result of doing this is a fraction, which of course could also be represented as a decimal or percent. This is best explained by example.

Imagine that we will roll a die once, and we're hoping that it will land on 3. Perhaps we will win a game should that occur. We can consider the die roll to be the event in question. There is one favorable outcome, namely a 3, and there are 6 possible outcomes, namely 1 through 6. We can write *P(3) = 1/6*. The context of the problem would have to make clear that that we are referring to the result of a die roll. The P(3) notation is read as, "The probability of 3," and the right side of the equation is read as "one-sixth" or "one out of six."

$$P(E) = \frac{\text{Number of favorable outcomes}}{\text{Number of total outcomes}}$$

The general formula for the probability of an event E

As another example, if we're going to draw a marble from a bag that contains 3 red marbles and 5 blue ones, the probability of drawing a blue marble is 5/8. There are 5 favorable outcomes out of a possible 8. We could write *P(blue) = 5/8*, but again, the problem would have to explain the context and all of the conditions.

THE CHANCE OF SOMETHING NOT HAPPENING

We can determine the chance of an event **not** happening by computing 1 minus the chance of the event happening. Recall that if we are dealing with percents, we can express the number 1 as 100%. That means if there is a 65% chance that it will rain, there is a 35% chance that it will not rain, computed as 100% – 65%. If we are dealing with fractions, we will use the value 1 instead of 100%.

For example, we saw that there is a 1/6 chance that an individual die roll will land on 3. To compute the probability of the die roll not landing on 3, we must compute 1 – 1/6. We should rewrite 1 as a fraction by making it 1/1. Multiply top and bottom by the LCD of 6, giving us a problem of 6/6 - 1/6 for an answer of 5/6. This makes sense intuitively. If there is a 1 in 6 chance to roll a 3, there is a 5 in 6 chance to roll something other than a 3.

CULTURAL ASPECTS OF PROBABILITY PROBLEMS

Some students feel that some probability problems are culturally biased because the problems make reference to things that they may not be familiar with such as a deck of playing cards. While this may be a valid concern, the situation is remedied with some very brief instruction.

We define a coin as a disk that is fairly flipped in the air. It has an equal chance of landing on one side which we call "heads," and the other side which we call "tails."

If a problem makes reference to a die, it is referring to a cube of six sides which is rolled, and which has an equal chance of landing on each side. The sides are numbered from 1 to 6, either using numerals or pips (spots). Some textbooks refer to dice as "number cubes" in order to avoid accusations of promoting gambling.

If a problem makes reference to drawing marbles or anything else out of any sort of container (often an urn), it is assumed that each such object has an equal chance of being drawn, and that there are no biases of any kind.

Many probability problems involve a standard deck of 52 playing cards. Each card is characterized by one of 4 suits, and one of 13 ranks. There are no duplicate cards so there are 4 × 13 or 52 different combinations. The suits are hearts, clubs, spades, and diamonds. The ranks run from Ace (effectively 1) through 10, followed by Jack, Queen, and King (effectively 11, 12, and 13, respectfully). Examples of cards are the "Seven of Diamonds" and the "King of Spades." Two suits are red and two are black which implies that half the deck is red and half is black.

TRICK QUESTIONS AND PROBABILITY MYTHS

There are many trick questions in probability that are easily answered as long as you don't fall victim to any common probability "myths." The most important concept is that each repetition of a probability experiment is completely and totally independent. For example, it doesn't matter how many times in a row a coin lands on heads. The chances of any given flip landing on heads is 50%. The coin is not "overdue" to land on tails, nor can it be said that heads are "on a hot streak" and are likely to continue. If you refuse to accept this you will get many easy test questions wrong.

Many probability questions include the word "fair" to remind the reader that no "tricks" are involved. Even if the word "fair" is omitted, we always assume total fairness and randomness. Decks of cards are assumed to be thoroughly shuffled. Die rolls and coin flips are performed such that no outcome is more likely than another. A problem will never deal with a "trick" coin or "weighted" die or any such thing. A problem will also never involve any unusual circumstance such as a coin landing on its edge or anything similar.

THE PROBABILITY OF COMPOUND INDEPENDENT EVENTS

Many probability problems ask us to compute the probability of two **independent** events both occurring. We define events as independent if they have nothing to do with each other. An example would be a coin toss followed by a die roll. Another example would be two coin tosses in a row, remembering what was just discussed about probability myths.

The simple rule to remember is that to determine the probability of multiple independent events occurring, we *multiply* the probabilities of the individual events. For example, to compute the probability of flipping a coin and having it land on heads, followed by rolling a die and having it land on 3, we would multiply 1/2 times 1/6 which are the probabilities of those independent events. We get an answer of 1/12 which is the probability of the two events occurring in tandem.

PROBABILITY WITH AND WITHOUT REPLACEMENT

Some probability problems include the phrase "with **replacement**" or "without replacement." This typically has to do with problems that involve drawing a card from a deck of cards, or drawing a marble out of a container. The question at hand is whether or not we will put back the card or marble prior to repeating the experiment. Obviously it makes a big difference.

For example, two marbles are drawn without replacement from an urn containing 4 red and 6 blue marbles. What is the probability that both marbles will be red?

We are essentially dealing with two independent events—a draw and a draw. We will multiply the individual probabilities like we just learned. The probability of drawing a red marble on the first draw is 4/10.

Think about the situation as it stands now. Since the drawn red marble was not replaced, the urn now has 3 red and 6 blue marbles. Now the chances of drawing a red are 3/9. Multiply the individual probabilities of 4/10 and 3/9 to get 12/90—the probability of drawing two red marbles in a row without replacement. Note that we often do not reduce probability fractions to lowest terms, but it is not wrong to do so.

Now let's do the same problem with the assumption that the problem stipulated the terms "with replacement." Now we are dealing with a probability of 4/10 on the first draw, followed once again by 4/10 on the second draw since the drawn red marble was returned to the urn. Again, multiply the two individual probabilities to get 16/100— the probability of drawing two red marbles in a row, with replacement.

PROBLEMS OF THE FORM "HOW MANY WAYS...?"

We are sometimes asked to solve problems involving the number of combinations that can be made by choosing different objects from different categories. These questions are extremely easy. Simply multiply the numbers involved, being careful to read the problem carefully to ensure that you are not being tricked in any way.

A typical question is as follows: "Jane is deciding what to wear. She is going to create an outfit by choosing a blouse, skirt, and hat from 5 different blouses, 4 different skirts, and 3 different hats. How many different outfits can she make?" All we do is multiply the given numbers. We compute $5 \times 4 \times 3$ to get 60. Almost all questions on this topic are as simple as that.

Certainly don't introduce any complications such as whether certain combinations should be disregarded because they involve components which unfashionably "clash." Just multiply the given numbers, and don't get side-tracked by any unrelated, extraneous information or numbers which may be included in the problem.

SO NOW WHAT?

The study of probability certainly gets much more involved than what has been presented. However, the vast majority of standardized test questions can be answered using nothing but the simple concepts explained in this chapter. Make sure that you fully understand them. You will probably even be able to answer many questions intuitively as long as you don't fall for any of the tricks or myths described in the chapter.

Be certain to always read probability questions slowly and carefully. It is often the case that misreading one word can totally change the entire problem, and of course most probability questions do come in the form of word problems.

Realistically, it is not worth spending all that much time on this chapter. Most students do not have much difficulty with probably questions, and if anything, have been known to find them "fun" or "interesting."

By far, your time is better spent ensuring that you are fully comfortable with the material on basic arithmetic, as well as fractions, decimals, and percents. You will work with those topics again and again as you progress to more advanced math such as algebra. If you don't master those topics now, you will simply have to master them later when you are busy with other work.

Be sure to read the next chapter which offers some tips on how to study and prepare for math exams.

CHAPTER TWELVE

How to Study and Learn Math, and Improve Scores on Exams

HOW *NOT* TO STUDY FOR A MATH TEST

Most students study so inefficiently for their math tests that it is worth beginning this section with a discussion of how *not* to study. Even if you do nothing more than avoid these things it will be beneficial.

First, understand that studying is not characterized by flipping through your notebook or textbook while nodding your head and saying, "Yeah, I know this stuff." By doing that, you are only confirming the fact that you remember having covered the given material at some point in the past. It means that you recognize it. It does not imply that you actually know it. If you want to study so that you will be ready for your exam, expect to do much more than just flipping through pages.

You will also accomplish very little by repeatedly performing (essentially copying) the same sample exercises from your book or notes over and over again. Almost certainly the problems on your exam will involve different numbers, and perhaps even different types of numbers such as negatives or decimals when perhaps you were only prepared for positive integers.

Your exam will probably also include variants of the sample problems which in some cases are the result of changing just one single word in the problem or instructions. If all you do is try to "spit back" what you learned by rote, you will likely get those questions wrong.

WHY STUDENTS DON'T STUDY EFFECTIVELY

As described in Chapter One, it is rare for teachers to provide actual instruction or guidance as to how to effectively study. The result is that most students interpret the assignment to mean that they should study in the manner described above.

It is not the fault of students that they typically "study" in this way. To do anything else requires either specific instruction and guidance, or a level of maturity and experience which may not yet have been attained.

IS IT EVEN NECESSARY TO STUDY FOR AN EXAM?

It is important to understand that the very concept of studying for exams is somewhat flawed. The only truly effective way to prepare for exams is to do so as you learn new material. This could be thought of as "studying as you go." Students who are disciplined enough and have the foresight for this need not even do any "studying" in the day(s) prior to an exam. In fact such students will probably just use those days to get themselves into the proper mindset to take their exam.

HOW TO LEARN MATH AND "STUDY AS YOU GO"

If you are taking a math class in school, it is important that you approach your coursework "in reverse." Be thinking about your final exam from day one. Of course you should do the same in regards to any other important exam in your future such as the SAT or similar, as well as your regular classroom exams. If you are preparing for an exam on your own such as the GED or a career-based exam, think about the exam as you progress through your studies. You will eventually be tested on everything that you are studying, so you might as well learn the material as you study it rather than pushing it aside with the logic that any test on it is far in the future.

It is important to resist the temptation to think of exams as hypothetical events in your future which may never come to pass. Unless something highly unexpected occurs in your life, tests will come your way. The more you think about them and prepare for them in advance, the less of a concern they'll be when they finally arrive.

PREPARING FOR "WHAT-IF" SCENARIOS

As you study for your exams, which again you should not do at the last minute, do your best to outguess the teacher or the test-maker. For example, if you were taught how to add fractions, and you were also taught about negative numbers, consider that your teacher may very well throw in a question about adding negative fractions. Ask yourself, "What if the teacher puts that on the test?," and don't accept an answer of, "Well, I just hope that s/he doesn't, but if s/he does I'll get it wrong and probably still pass." Instead, just learn how to do the very type problem that you are concerned about.

Trying to "beat the test" is not at all foolproof, and certainly many tests include a few unique and unexpected problems that require original and creative thinking. However, you should certainly try to anticipate any types of problems that have a high likelihood of appearing on your exams, and make sure that you know how to solve them. If you do this in advance there will be plenty of time in which to seek extra help if needed.

When studying for a test, try to think of as many variants of problems as you can, being a creative as possible. For example, are you going to get flustered if you see a fraction that involves decimal values? If you've learned about both fractions and decimals, then such a thing is very much "fair game." Make it a point to go out of your way to try to trick yourself, as unpleasant as that may sound. In so doing, you will be better prepared for any "trick questions" on the part of the teacher or test-maker.

Make sure you're prepared for questions in a different order than the material was taught. Don't study with the assumption that the questions will be given in any particular order or grouped in any particular way such as by topic or difficulty level. Constantly ask yourself, "Is there anything that is going to confuse me, or anything that I simply don't know how to do?," and then remedy the situation so that it won't be a concern.

AVOIDING A "PASS OR FAIL" MINDSET

Unfortunately, most students believe in a pass/fail model of grading. Passing is taken to mean that the student "did good," and gets to follow his/her friends into the next grade. Failing is taken to mean that the student "did bad," and that s/he must retake the class in summer school instead of being able to go outside and play. It also means that if the class isn't passed in summer school, it will have to be retaken again the following year with all new (and younger) classmates.

This mindset typically results in failure, or at best, passing "by the skin of one's teeth." The simple reason is that the student has effectively set his/her target as the minimal passing grade. Such a student hates to earn a much higher grade because that would imply that s/he "over-studied" and wasted his/her time.

To do well in math you must "aim high." Your goal should be to score high on exams, and to do so easily and with plenty of time to spare. It doesn't matter whether or not you actually achieve that goal, or how unrealistic you think it may be. What is important is cultivating that mindset, and doing so from the first day of school each year long before any exams are announced.

By no means should you create stress for yourself in the process, nor should you be disappointed if you don't actually attain your goal. The point is to avoid a self-fulfilling a negative or pessimistic prophecy. Visualize yourself doing well on exams. Abandon the mindset of hoping to just barely pass. By doing this you have nothing to lose and everything to gain. Having the right mindset won't miraculously implant math knowledge in your head, but it will mean that you at least have the mental preparation to take what you know and put it to best use. It will also likely result in your studying harder throughout the year in preparation for your exams.

PRACTICING MEDITATION TO CULTIVATE THE OPTIMAL MINDSET FOR EXAMS

Many students know the math that they are being tested on, but make careless mistakes on their exams due to nervousness or lack of focus. Sometimes these feelings can be so out of control that the student is unable to even look at the first question. An excellent way to counteract this is through the practice of meditation.

To meditate literally means to just maintain a single point of focus. While it common for the object of that focus to be the workings of one's own mind, it could really be anything at all. The goal is simply to cultivate the ability to focus our attention while keeping any mental "drifting" to a minimum. The goal is furthermore to become aware of when such is occurring, and to be able to gently bring our minds back to the task at hand.

This is certainly the mindset which is needed during exam conditions. When we are nervous it is especially easy for our minds to start running wild, often unbeknownst to us until the bell rings, and the teacher instructs us to put down our pencils and hand in our exam.

If you want to practice meditation, just sit comfortably in a quiet room with no distractions. The goal is to be relaxed, yet alert, which means you should avoid sitting stiffly or in any posture which will make it difficult to maintain focus. Breathe naturally, watching your breath to help bring yourself into a focused state of mind.

Once your thoughts have settled down to some extent, simply pick something that can serve as the focus of your attention. It could be something such as counting your breaths to 4, and then restarting at 1. It could be a very short sequence of meaningful words. It could be an object of some kind. It could even be your own mind although once you get a glimpse of what is taking place in there, that can lead to its own points of concern.

The point of the exercise is to practice being focused. If you lose focus, just gently bring yourself back. Don't worry about how long you were out of focus, or what caused you to get out of it. There is no room for analysis during your meditation practice. If you're afraid that you'll lose track of time, you can set a timer for yourself, or you can just stop whenever you choose to stop.

With regular practice, you'll find that you'll be able to take your power of focus and carry it over into your everyday tasks outside your meditation sessions. If you lose your focus during an exam, you will have had practice with how to be aware of such, and how to bring it back. Meditation helps us build our sense of alertness and awareness, while simultaneously bringing us into a state of mental calmness. It is the absolute ideal state of mind with which to take a math exam.

DEALING WITH ANXIETY BEFORE / DURING A TEST

A very simple and effective technique to combat exam-related anxiety is to simply keep breathing and being aware of your breath. If you take careful notice of yourself the next time you are feeling nervous or overwhelmed in any context, you will probably observe that your breathing has become shallow and rapid.

In any case, it is important to be aware of your breath at all times, especially before and during exams. Certainly don't place all of your focus on your breathing because then you won't be able to focus on the exam itself, but just devote some portion of your brain to maintaining a steady breathing pattern. When this is challenging, simply look away from the exam, take a slow deep breath, and gently bring yourself back into focus. The short time that you lose will result in the remainder of your exam time being that much more productive.

MORE PRACTICAL TIPS FOR REDUCING ANXIETY

Of course there are many practical steps that can be done to minimize exam anxiety. As discussed, if you over-learn the material instead of just aiming for a 65, you are less likely to look at your exams and panic. Instead, you'll have a big smile on your face knowing that you can leisurely take half the time to complete the exam, and use the rest of the time to stare out the window.

Never get hung up on any one particular question. While many exams do begin with an easy question to serve as a "warm-up," some exams can do the exact opposite. If you're finding yourself getting overwhelmed by a question after giving it a bit of effort, simply mark it so that you'll know to get back to it later, and then move on. Do not assume that the entire test will be just as hard as whatever question you set aside, or that the other questions will even be on the same topic.

Once you are allowed to begin working on your test, use the first minute of your time to simply relax, breathe, and get your bearings. Many students attack their exams with the logic that anything else will result in wasted time. Having one less minute of time won't negatively alter your grade, and almost certainly will greatly raise it since you'll take your exam with a sense of mindfulness.

Make it a point to avoid interacting with other students before exams, especially students who are bragging about how easy the test will be for them, or expressing concern that they will fail, or any similar negativity. Certainly don't try to pick up any last minute facts or tips from your classmates since this will likely just confuse and fluster you more than it will benefit you.

Use the time before an exam to breathe, and cultivate the sense of mindfulness that comes through the regular practice of meditation. It is also probably best that you not think about the actual test material until the exam begins. You're not going to forget everything if you stop reciting it to yourself over and over again like most students do before an exam. Exams are designed to determine the extent to which you have internalized the material. Your performance will be optimal if you simply relax while being alert and focused.

LEARNING TO FOLLOW INSTRUCTIONS

Many students submit an exam convinced that they have scored very high, only to receive it back with a low grade due to a lack of following instructions. As obvious as this sounds, you must develop the discipline to read all exam instructions carefully, and carry them out. This applies to instructions for individual problems as well as instructions that apply to the exam itself such as how to or where to write your answers. This is something that cannot be taught, but simply must be practiced. It is also the most important skill from school that you will use in your everyday life, especially your career.

CHECKING ANSWERS FOR REASONABLENESS

Many careless errors can be caught and corrected by simply learning how to check your answers for reasonableness. This is especially applicable to word problems. Before even beginning a problem, see if you can estimate the answer. For some problems this will be easier than others. Once you obtain an answer for a problem, don't just write it down and forget about it, and certainly don't rationalize doing this by saying that you're just writing what your calculator told you. See if your answer makes some sense with the context of the problem. For example, a problem involving the distance that someone walked will probably have an answer between 0 and 20 miles. Any answer outside that range is almost certainly wrong.

CAN YOU TEACH THE TOPICS TO SOMEONE ELSE?

The best way to know if you are prepared for an exam is to see if you can teach the topics to someone else. If you don't have anyone to practice this with, you can even just teach a fictitious person in your head. It doesn't matter how good your teaching skills are. The point of this exercise is to see if you are confident enough to actually convey the information that you have learned, which is exactly what you will be asked to do on your exam. If you cannot teach a topic to your real or fictitious study partner because you haven't truly mastered it yourself, take the time to do so. The further in advance you begin studying, the easier this will be.

BASIC LOGISTICAL ISSUES OF TAKING EXAMS

As obvious as this sounds, don't lose any points as a result of simple, easily avoidable logistical issues. Make sure that you have whatever materials will be needed for your exam such as a calculator and pens or pencils. Don't waste time on your exam by begging your teacher for hints or any special help that you are not entitled to.

Don't argue the "fairness" of any question which is something that can be done after the exam, if at all. Be sure that your answers are neat, and that you have shown all your work if required. Don't cheat in any way, or do anything that could be construed as cheating such as making sounds, or fidgeting, or digging in your bag.

THE FINAL WORD

The fact that you have reached this paragraph confirms many things about you. First, it means that you're a good reader. This is actually a prerequisite for math success since instructions and word problems must be read very carefully. If you weren't a good reader, you probably would have stopped reading this book long before now.

It also means that you have a sense of perseverance. Math is a dry and challenging subject for most, and yet you have stuck with it. That is a personally trait which is found in students who succeed in math.

Just the fact that you even purchased this book proves the fact that you want to achieve your math goals, and that at least at some level, you believe that you can. You had the choice to either buy no book at all, or buy a commercial review book, and yet somehow you were drawn to buying and reading this book cover to cover. As mentioned, we tend to fulfill our own prophecies. If you believe that math can be easier, then that is precisely what it will become.

As explained in Chapter One, I truly believe that virtually everyone can succeed in their math goals. The small percentage of the population that either can't or refuse to are not among the purchasers or readers of this book.

Since you already own this book, make good use of it. Reread it as necessary to make sure that you fully understand all of the concepts that were presented. Reread this chapter on how to study for exams.

You can do it! You can achieve whatever math goals you have set for yourself, but doing so will certainly take time and effort. In fact, it will take precisely the amount of time and effort that is necessary—no more and no less.

Contact me via my website if you have questions about the material or would like to discuss your academic situation. Study hard and believe in yourself! ☺

About the Author

Larry Zafran was born and raised in Queens, NY where he tutored and taught math in public and private schools. He has a Bachelors Degree in Computer Science from Queens College where he graduated with highest honors, and has earned most of the credits toward a Masters Degree in Secondary Math Education.

He is a dedicated student of the piano, and the leader of a large and active group of board game players which focuses on abstract strategy games from Europe.

He presently lives in Cary, NC where he works as an independent math tutor, writer, and webmaster.

Companion Website for More Help

For free support related to this or any of the author's math books, please visit the companion website below.

www.MathWithLarry.com

Made in the USA
Lexington, KY
23 December 2009